The Open
University

The
Molecular World

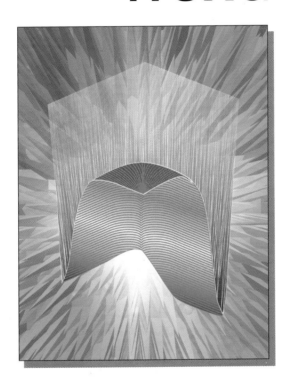

Chemical Kinetics and Mechanism

edited by

Michael Mortimer and Peter Taylor

This publication forms part of an Open University course, S205 *The Molecular World*. Most of the texts which make up this course are shown opposite. Details of this and other Open University courses can be obtained from the Call Centre, PO Box 724, The Open University, Milton Keynes MK7 6ZS, United Kingdom: tel. +44 (0)1908 653231, e-mail ces-gen@open.ac.uk

Alternatively, you may visit the Open University website at http://www.open.ac.uk where you can learn more about the wide range of courses and packs offered at all levels by The Open University.

The Open University, Walton Hall, Milton Keynes, MK7 6AA

First published 2002

Edited, designed and typeset by The Open University.

Published by the Royal Society of Chemistry, Thomas Graham House, Science Park, Milton Road, Cambridge CB4 0WF, UK.

Printed in the United Kingdom by Bath Press Colourbooks, Glasgow.

ISBN 0 85404 670 4

A catalogue record for this book is available from the British Library.

1.1

s205book 5 i1.1

The Molecular World

This series provides a broad foundation in chemistry, introducing its fundamental ideas, principles and techniques, and also demonstrating the central role of chemistry in science and the importance of a molecular approach in biology and the Earth sciences. Each title is attractively presented and illustrated in full colour.

The Molecular World aims to develop an integrated approach, with major themes and concepts in organic, inorganic and physical chemistry, set in the context of chemistry as a whole. The examples given illustrate both the application of chemistry in the natural world and its importance in industry. Case studies, written by acknowledged experts in the field, are used to show how chemistry impinges on topics of social and scientific interest, such as polymers, batteries, catalysis, liquid crystals and forensic science. Interactive multimedia CD-ROMs are included throughout, covering a range of topics such as molecular structures, reaction sequences, spectra and molecular modelling. Electronic questions facilitating revision/consolidation are also used.

The series has been devised as the course material for the Open University Course S205 *The Molecular World*. Details of this and other Open University courses can be obtained from the Course Information and Advice Centre, PO Box 724, The Open University, Milton Keynes MK7 6ZS, UK; Tel +44 (0)1908 653231; e-mail: ces-gen@open.ac.uk. Alternatively, the website at www.open.ac.uk gives more information about the wide range of courses and packs offered at all levels by The Open University.

Further information about this series is available at www.rsc.org/molecularworld.

Orders and enquiries should be sent to:

Sales and Customer Care Department, Royal Society of Chemistry, Thomas Graham House, Science Park, Milton Road, Cambridge, CB4 0WF, UK

Tel: +44 (0)1223 432360; Fax: +44 (0)1223 426017; e-mail: sales@rsc.org

The titles in *The Molecular World* series are:

THE THIRD DIMENSION
  edited by Lesley Smart and Michael Gagan

METALS AND CHEMICAL CHANGE
  edited by David Johnson

CHEMICAL KINETICS AND MECHANISM
  edited by Michael Mortimer and Peter Taylor

MOLECULAR MODELLING AND BONDING
  edited by Elaine Moore

ALKENES AND AROMATICS
  edited by Peter Taylor and Michael Gagan

SEPARATION, PURIFICATION AND IDENTIFICATION
  edited by Lesley Smart

ELEMENTS OF THE p BLOCK
  edited by Charles Harding, David Johnson and Rob Janes

MECHANISM AND SYNTHESIS
  edited by Peter Taylor

The Molecular World Course Team

Course Team Chair
Lesley Smart

Open University Authors
Eleanor Crabb (Book 8)
Michael Gagan (Book 3 and Book 7)
Charles Harding (Book 9)
Rob Janes (Book 9)
David Johnson (Book 2, Book 4 and Book 9)
Elaine Moore (Book 6)
Michael Mortimer (Book 5)
Lesley Smart (Book 1, Book 3 and Book 8)
Peter Taylor (Book 5, Book 7 and Book 10)
Judy Thomas (*Study File*)
Ruth Williams (skills, assessment questions)

Other authors whose previous contributions to the earlier courses S246 and S247 have been invaluable in the preparation of this course: Tim Allott, Alan Bassindale, Stuart Bennett, Keith Bolton, John Coyle, John Emsley, Jim Iley, Ray Jones, Joan Mason, Peter Morrod, Jane Nelson, Malcolm Rose, Richard Taylor, Kiki Warr.

Course Manager
Mike Bullivant

Course Team Assistant
Debbie Gingell

Course Editors
Ian Nuttall
Bina Sharma
Dick Sharp
Peter Twomey

CD-ROM Production
Andrew Bertie
Greg Black
Matthew Brown
Philip Butcher
Chris Denham
Spencer Harben
Peter Mitton
David Palmer

BBC
Rosalind Bain
Stephen Haggard
Melanie Heath
Darren Wycherley
Tim Martin
Jessica Barrington

Course Reader
Cliff Ludman

Course Assessor
Professor Eddie Abel, University of Exeter

Audio and Audiovisual recording
Kirsten Hintner
Andrew Rix

Design
Steve Best
Debbie Crouch
Carl Gibbard
Sara Hack
Sarah Hofton
Mike Levers
Sian Lewis
Jenny Nockles
John Taylor
Howie Twiner

Library
Judy Thomas

Picture Researchers
Lydia Eaton
Deana Plummer

Technical Assistance
Brandon Cook
Pravin Patel

Consultant Authors
Ronald Dell (*Case Study:* Batteries and Fuel Cells)
Adrian Dobbs (Book 8 and Book 10)
Chris Falshaw (Book 10)
Andrew Galwey (*Case Study:* Acid Rain)
Guy Grant (*Case Study:* Molecular Modelling)
Alan Heaton (*Case Study:* Industrial Organic Chemistry, *Case Study:* Industrial Inorganic Chemistry)
Bob Hill (*Case Study:* Polymers and Gels)
Roger Hill (Book 10)
Anya Hunt (*Case Study:* Forensic Science)
Corrie Imrie (*Case Study:* Liquid Crystals)
Clive McKee (Book 5)
Bob Murray (*Study File*, Book 11)
Andrew Platt (*Case Study:* Forensic Science)
Ray Wallace (*Study File*, Book 11)
Craig Williams (*Case Study:* Zeolites)

CONTENTS

PART 1 CHEMICAL KINETICS

Clive McKee and Michael Mortimer

USE OF THE CD-ROM PROGRAM: *KINETICS TOOLKIT*		10
1	INTRODUCTION	11
	1.1 A general definition of rate	14
2	A CLOSER LOOK AT CHEMICAL REACTIONS	18
	2.1 Individual steps	19
	2.2 Summary of Section 2	23
3	RATE IN CHEMICAL KINETICS	24
	3.1 Kinetic reaction profiles	24
	3.2 Rate of change of concentration of a reactant or product with time	27
	3.3 A general definition of the rate of a chemical reaction	31
	3.4 Summary of Section 3	33
4	FACTORS DETERMINING THE RATE OF A CHEMICAL REACTION	35
	4.1 A simple collision model	35
	4.2 An experimental approach	38
	4.3 Summary of Section 4	40
5	DETERMINING EXPERIMENTAL RATE EQUATIONS AT A FIXED TEMPERATURE	42
	5.1 Practical matters	42
	5.2 A strategy	43
	5.3 Reactions involving a single reactant	44
	5.3.1 A preliminary half-life check	44
	5.3.2 The differential method	47
	5.3.3 The integration method	52

5.4	**Reactions involving several reactants**	**57**
	5.4.1 The isolation method	57
	5.4.2 The initial rate method	60
5.5	**Summary of Section 5**	**63**

6	THE EFFECT OF TEMPERATURE ON THE RATE OF A CHEMICAL REACTION	65
6.1	**The Arrhenius equation**	**65**
6.2	**Determining the Arrhenius parameters**	**67**
6.3	**The magnitude of the activation energy**	**73**
6.4	**Summary of Section 6**	**78**

7	ELEMENTARY REACTIONS	80
7.1	**Molecularity and order**	**80**
7.2	**Reactions in the gas phase**	**81**
7.3	**Reactions in solution**	**85**
7.4	**Femtochemistry**	**86**
7.5	**Summary of Section 7**	**91**

8	REACTION MECHANISM	92
8.1	**Evidence that a reaction is composite**	**92**
8.2	**A procedure for simplification: rate-limiting steps and pre-equilibria**	**94**
8.3	**Confirmation of a mechanism**	**100**
8.4	**Summary of Section 8**	**103**

SUMMARY OF PART 1	104
LEARNING OUTCOMES FOR PART 1	105
QUESTIONS: ANSWERS AND COMMENTS	108
EXERCISES: ANSWERS AND COMMENTS	120
FURTHER READING	131
ACKNOWLEDGEMENTS	132

PART 2: THE MECHANISM OF SUBSTITUTION

Edited by Peter Taylor from work authored by Richard Taylor

1	ORGANIC REACTIONS	135
	1.1 Why are organic reactions important?	135
	1.2 Classification of organic reactions	136
2	REACTION MECHANISMS	139
	2.1 Reaction mechanisms: why study them?	139
	2.2 Breaking and making covalent bonds	144
	2.2.1 Radical reactions	145
	2.2.2 Ionic reactions	147
	2.3 Summary of Sections 1 and 2	148
3	IONIC SUBSTITUTION REACTIONS	150
	3.1 Nucleophiles, electrophiles and leaving groups	150
	3.2 The scope of the S_N reaction	154
	3.2.1 Nucleophiles	154
	3.2.2 Leaving groups	156
	3.3 How far and how fast?	161
	3.3.1 How far?	161
	3.3.2 How fast?	163
	3.4 Summary of Section 3	164
4	S_N2 AND S_N1 REACTION MECHANISMS	166
	4.1 Introduction	166
	4.2 Kinetics and mechanism of S_N reactions	167
	4.2.1 A concerted mechanism	168
	4.2.2 Two-step associative mechanism	168
	4.2.3 Two-step dissociative mechanism	169
	4.2.4 Which mechanism is at work?	171
	4.3 Summary of Section 4	174
5	S_N2 VERSUS S_N1	175
	5.1 The effect of substrate structure	175
	5.2 The effect of the nucleophile	177
	5.3 Summary of Section 5	178
6	CONCLUDING REMARKS	179
	LEARNING OUTCOMES FOR PART 2	180
	QUESTIONS: ANSWERS AND COMMENTS	182
	ACKNOWLEDGEMENTS	185

PART 3: ELIMINATION: PATHWAYS AND PRODUCTS

Edited by Peter Taylor from work authored by Richard Taylor

1	INTRODUCTION: β-ELIMINATION REACTIONS	189
	1.1 The mechanisms of β-elimination reactions	191
	1.2 Summary of Section 1	193
2	THE E2 MECHANISM	194
	2.1 The scope of the E2 mechanism	194
	2.2 The stereochemistry of the E2 mechanism	195
	2.3 Isomeric alkenes in E2 reactions	198
	2.3.1 Which isomer will predominate?	199
	2.3.2 Which direction of elimination?	200
3	THE E1 MECHANISM	203
	3.1 Summary of Sections 2 and 3	206
4	ELIMINATION VERSUS SUBSTITUTION	208
	4.1 Substrate structure	208
	4.1.1 Unimolecular versus bimolecular mechanism	209
	4.2 Choice of reagent and other factors	209
	4.2.1 Choice of leaving group	210
	4.2.2 Temperature	210
	4.2.3 Summing up	210
	4.3 Summary of Section 4	211
5	OTHER USEFUL ELIMINATION REACTIONS	212
	5.1 Dehalogenation and decarboxylative elimination	212
	5.2 Preparation of alkynes by elimination reactions	213
	5.3 Summary of Section 5	214
	LEARNING OUTCOMES FOR PART 3	215
	QUESTIONS: ANSWERS AND COMMENTS	216
	ACKNOWLEDGEMENTS	224

CASE STUDY: SHAPE-SELECTIVE CATALYSIS USING ZEOLITES

Craig Williams and Michael Gagan

1	INTRODUCTION		227
	1.1	Natural zeolites	227
	1.2	Synthetic zeolites	228
2	STRUCTURE, PROPERTIES AND CLASSIFICATION OF ZEOLITES		230
	2.1	Basic structures	230
	2.2	Zeolite properties	232
	2.3	Zeolites as catalysts	234
	2.4	Zeolite classification	235
	2.5	Small-pore zeolites	237
	2.6	Medium-pore zeolites	238
	2.7	Large-pore zeolites	239
3	SHAPE SELECTIVITY		242
	3.1	Mass-transport discrimination	242
	3.2	Transition-state selectivity	245
	3.3	Molecular traffic control	245
4	APPLICATIONS OF SHAPE SELECTIVITY		246
	4.1	*Para* selective alkylation of aromatic hydrocarbons	246
	4.2	Selective xylene isomerization	246
	4.3	Some other selective alkylation reactions of aromatic compounds	247
	4.4	Methanol to gasoline	248
5	ZEOLITES AS ENZYME MIMICS		250
6	MESOPOROUS ALUMINOSILICATE STRUCTURES		251
7	CONCLUSION		255
	ACKNOWLEDGEMENTS		256
	INDEX		257
	CD-ROM INFORMATION		262

Part 1

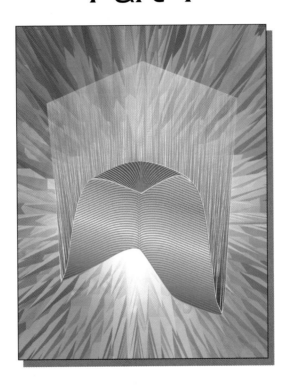

Chemical Kinetics

Clive McKee and Michael Mortimer

USE OF THE CD-ROM PROGRAM: *KINETICS TOOLKIT*

The CD-ROM program *Kinetics Toolkit* is an essential part of the main text. It is a graphical plotting application which allows data to be input, manipulated and then plotted. The plotted data can be analysed, for example to obtain the slope of a straight line. All data that are input can be stored in files for future use. Instructions for using the program are on the CD-ROM. Help files are available from the *Help* menus of the *Kinetics Toolkit*.

The *Kinetics Toolkit* is provided so that you can focus your attention on the underlying principles of the analysis of chemical kinetic data rather than becoming involved in the time-consuming process of manipulating data sets and graph plotting. Full sets of data are provided for most of the examples that are used in the main text and you should, as a matter of course, use the *Kinetics Toolkit* to follow the analysis that is provided. A number of the Questions, and *all* of the Exercises, require you to use the *Kinetics Toolkit* in answering them.

Ideally you should have direct access to your computer with the *Kinetics Toolkit* installed when you study Sections 1, 3, 5 and 6.

As a matter of priority you should try to do Exercise 1.1 in Section 1 as soon as possible since it is designed to introduce you to the use and scope of the *Kinetics Toolkit*.

It is still possible to study Sections 3, 5 and 6 if you are away from your computer, but you will need to return to those parts, including Questions and Exercises, that require the use of the *Kinetics Toolkit* at a later time.

A summary of the main use of the *Kinetics Toolkit* in Sections 3, 5, and 6, in the order of appearance in the text, is as follows:

Section 3:   Question 3.2, Exercise 3.1

Section 5:   Section 5.3.1, Section 5.3.2, Question 5.1, Exercise 5.1, Question 5.2, Exercise 5.2, Question 5.3

Section 6:   Section 6.2, Exercise 6.1

You may wish to use this summary both to plan your study and also act as a checklist. The most intensive use of the *Kinetics Toolkit* is in Section 5.

INTRODUCTION

Movement is a fundamental feature of the world we live in; it is also inextricably linked with time. The measurement of time relies on change — monitoring the swing of a pendulum, perhaps — but conversely, any discussion of the motion of the pendulum must involve the concept of time. Taken together, time and change lead to the idea of *rate*, the quantity which tells us how much change occurs in a given time. Thus, for example, for our pendulum we might describe rate in terms of the number of swings per minute. Or, to take a familiar example from everyday life, we refer to a rate of change in position as *speed* and measure it as the distance travelled in a given time (Figure 1.1).

The study of movement in general is the subject of *kinetics* and **chemical kinetics**, in particular, is concerned with the measurement and interpretation of the rates of chemical reactions. It is an area quite distinct from that of chemical thermodynamics which is concerned only with the initial states of the reactants (before a reaction begins) and the final state of the system when an *equilibrium* is reached (so that there is no longer any *net* change). What happens between these initial and final states of reaction and exactly how, and how quickly, the transition from one to the other occurs is the province of chemical kinetics. At the *molecular level* chemical kinetics seeks to describe the behaviour of molecules as they collide, are transformed into new species, and move apart again. But there is also a very practical side to the subject which is quickly appreciated when we realize that our very existence depends on a balance between the rates of a multitude of chemical processes: those controlling our bodies, those determining the growth of the animals and plants that we eat, and those influencing the nature of our environment. We must also not forget those changes that form the basis of much of modern technology, for which the car provides a wealth of examples (see Box 1.1). Whatever the process, however, information on how quickly it occurs and how it is affected by external factors is of key importance. Without such knowledge, for example, we would be less well-equipped to generate products in the chemical industry at an economically acceptable rate, or design appropriate drugs, or understand the processes that occur within our atmosphere.

Figure 1.1
The tennis ball may well leave the racket at a speed greater than 55 metres per second.

Historically, the first quantitative study of a chemical reaction is considered to have been carried out by Ludwig Wilhelmy in 1850. He followed the breakdown of sucrose (cane sugar) in acid solution to give glucose and fructose and noted that the rate of reaction at any time following the start of reaction was directly proportional to the amount of sucrose remaining unreacted at that time. For this observation Wilhelmy richly deserves to be called 'the founder of chemical kinetics'. Just over a decade later Marcellin Berthelot and Péan de St Gilles made a similar but more significant observation. In a study of the reaction between ethanoic acid (CH_3COOH) and ethanol (C_2H_5OH) to give ethyl ethanoate ($CH_3COOC_2H_5$) they found the measured rate of reaction at any instant to be approximately proportional to *the concentrations of the two reactants at that instant multiplied together*. At the time, the importance of this result was not appreciated but, as we shall see, relationships of this kind are now known to describe the rates of a wide range of different chemical processes. Indeed, such relationships lie at the heart of

BOX 1.1 Chemistry and cars

No matter what reasons influence the choice of a new car, it may not be immediately obvious that a considerable degree of faith is being placed in the rates of a number of rather complex chemical processes when we do so. On the one hand is a set of reactions which will be, one can only hope, as slow as possible. The photochemical processes causing the showroom brightness of paint and plastics to fade, and the attack by oxygen, moisture and road salt which leads to corrosion of metal parts, should have timescales of years, while we expect the deterioration of lubricating oil at high temperatures in the engine to be minimal during the period between services.

By contrast, there are other reactions which must be fast. When the ignition key is turned, an electrochemical process in the battery should quickly deliver a substantial electric current to turn the starter motor. In each cylinder of the engine, the air–fuel mixture should ignite at the optimum moment during the cycle of the piston, and burn rapidly and smoothly. If conditions are not exactly right, the chemical kinetics become erratic, combustion is accompanied by 'knock' (a series of disorderly explosions), engine efficiency falls, and the eventual result may be damage to the piston. When the exhaust gases leave the engine, but before their final discharge to the atmosphere, chemistry faces another challenge: clean air regulations must be met. This means that a catalytic converter has to be placed in the exhaust line which, during the 70 milliseconds or so taken by the gases to pass through it, must simultaneously and with very high efficiency, promote three different types of chemical reaction. Carbon monoxide (CO) is oxidized to carbon dioxide (CO_2), any unburnt fuel is oxidized to CO_2 and H_2O, and oxides of nitrogen which form in the cylinder from atmospheric N_2 and O_2 are reduced to nitrogen.

Lastly, it may seem somewhat bizarre to associate a safety device with an explosion but this is the method used to inflate air bags (Figure 1.2). If a vehicle is involved in a frontal collision equivalent to hitting a brick wall at more than 12 to 15 miles per hour, then sensors detect the sudden deceleration and trigger an electrical device which detonates a mixture containing about 65 g of sodium azide (NaN_3). The azide undergoes a relatively slow type of explosion

(technically, a deflagration), liberating approximately 35 litres of nitrogen gas

$$2NaN_3(s) = 2Na(s) + 3N_2(g)$$

The sodium formed is a potential hazard and is converted by reaction with other ingredients in the mixture (potassium nitrate, KNO_3 and silica, SiO_2) into an inert alkaline silicate 'glass'.

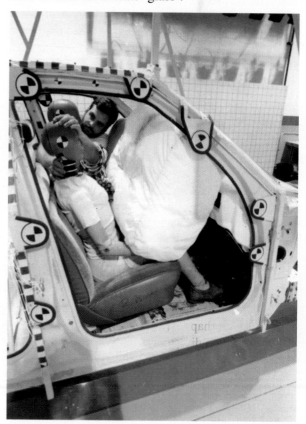

Figure 1.2 A test of air bag inflation.

To ensure maximum protection it is vital that the air bag, which is initially expanding very rapidly, is fully inflated within about 30 milliseconds. This then leaves a short period of time, 20 milliseconds or so, before the occupant hits it. Furthermore, by the time of impact, the nitrogen gas which has emerged from the explosion at a relatively high temperature is already cooling and, as a consequence, its volume is decreasing. Overall, at impact the airbag is deflating and this, along with the loss of gas through vents, provides a more effective cushion against serious injury.

empirical chemical kinetics, that is an approach to chemical kinetics in which the aim is to describe the progress of a chemical reaction with time in the simplest possible mathematical way.

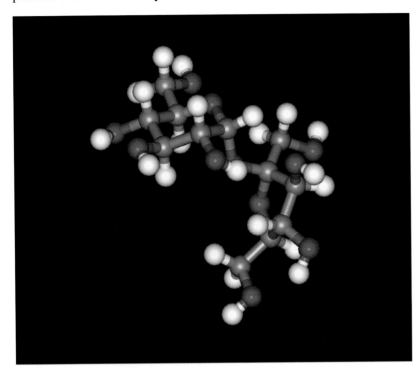

Figure 1.3
Sucrose is a carbohydrate in which two sugar or monosaccharide units, each with a particular ring structure, are linked together to give a disaccharide. In acid solution, the link is broken (hydrolysis) and the ring structures separate to yield glucose and fructose. (In the chemical structure shown, the hydrogen atoms attached directly to the individual carbon atoms in the two rings have been omitted in order to give a clearer view of the overall shape of the molecule.) * 🖥

By the 1880s, the study of reaction rates had developed sufficiently to be recognized as a discipline in its own right. The 21 December 1882, issue of the journal *Nature* noted,

> 'What may perhaps be called the kinetic theory of chemical actions, the theory namely, that the direction and amount of any chemical change is conditioned not only by the affinities, but also by the masses of reacting substances, by the temperature, pressure, and other physical circumstances — is being gradually accepted, and illustrated by experimental results.'

Over a century later, chemical kinetics remains a field of very considerable activity and development; indeed nine Nobel prizes in Chemistry have been awarded in this subject area. The most recent (1999) was to A. H. Zewail whose work revealed for the first time what actually happens at the moment in which chemical bonds in a reactant molecule break and new ones form to create products. This gives rise to a new area: *femtochemistry*. The prefix femto (abbreviation 'f') represents the factor 10^{-15} and indicates the timescale, which is measured in femtoseconds, of the new experiments. As some measure of how short a femtosecond is, while you read these words light is taking about 2 million femtoseconds (2×10^6 fs) to travel from the page to your eye and a further 1 000 fs to pass through the lens to the retina.

* This symbol, 🖥, indicates that this Figure is available in WebLab ViewerLite on the CD-ROM associated with this book.

BOX 1.2 Chemistry in bones: a very slow reaction

Chemical change can be observed over a vast range of timescales, from femtoseconds to thousands of years. An archaeological investigation provides an example of one of the slowest reactions ever studied quantitatively. The naturally occurring amino acids (with the exception of glycine) have a chiral carbon atom, and so the molecules can exist in two enantiomeric forms. Only the *S*-configuration is usually found in *living* tissue. However, *S*-molecules can slowly undergo *inversion* to the mirror-image *R*-form and once an organism dies the *R*:*S* ratio in bone, say, will increase slowly with time. In the case of aspartic acid (Figure 1.4), this increase occurs on a timescale of thousands of years and, as a consequence, measurement of the *R*:*S* aspartic acid ratio provides a convenient means of dating human or animal fossil remains.

This dating method was used in 1974 to determine the age of hominid bones found in California. The results indicated that humans may have been present in the region for up to 50 000 years. This puts into question the previous assumption that the New World was first populated by migration from Asia around 15 000 to 25 000 years ago. This is the most recent time during which a land bridge existed where the Bering Straits now separate Russia and Alaska. The geological record, however, shows that sea-levels were sufficiently low for a similar bridge to have existed about 70 000 years ago and so it would appear that the original migration may have occurred during that much earlier period.

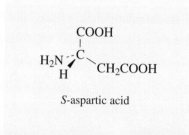

S-aspartic acid

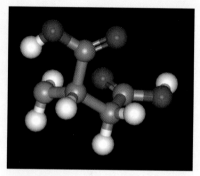

Figure 1.4
S-aspartic acid.

QUESTION 1.1

In an empirical approach to chemical kinetics, what would be the simplest mathematical way of representing the information obtained by Marcellin Berthelot and Péan de St Gilles for the reaction between ethanoic acid and ethanol?

1.1 A general definition of rate

So far, we have tended to use the term *rate* in a purely qualitative way. However, it is important for later discussions to introduce a more quantitative definition. In one sense, rate is the amount of one thing which corresponds to a certain amount (usually one) of some other thing. For example, governments, financial markets and holidaymakers in foreign countries may be concerned about exchange rates: the number of dollars, euros or other currency that can be bought for one pound sterling. More frequently, however, and as we have mentioned earlier, the concept of rate involves the passage of time. This is particularly so in the area of chemical kinetics. We shall restrict our definitions of rate, therefore, to cases in which time is involved.

For a physical quantity that changes *linearly* with time, we can take as a definition:

$$\text{rate of change} = \frac{\text{change in physical quantity in typical units}}{\text{time interval in typical units}} \qquad (1.1)$$

STUDY NOTE

What follows is a summary of key ideas relating to the definition of rate in cases in which time is involved. Exercise 1.1 at the end of this section provides you with an opportunity to test your understanding of rate of change.

For time, typical units are seconds, minutes, hours, and so on. If, for example, the physical quantity was distance then typical units could be metres and the rate of change would correspond to speed measured in, say, metres per second ($m\ s^{-1}$). Since the physical quantity changes linearly with time this means that the change in any one time interval is exactly the same as that in any other equal interval. In other words a plot of physical quantity versus time will be a *straight line* and there is a uniform, or *constant*, rate of change.

Equation 1.1 can be written in a more compact notation. If the physical quantity is represented by y, then it will change by an amount Δy during a time interval Δt, and we can write

$$\text{rate of change of } y \ = \ \frac{\Delta y}{\Delta t} \tag{1.2}$$

This rate of change, $\Delta y/\Delta t$, corresponds mathematically to the *slope* (or *gradient*) of the straight line and, as already stated, has a constant value.

A very important situation arises when a rate of change itself varies with time. A familiar example is a car accelerating; as time progresses, the car goes faster and faster. In this case a plot of physical quantity versus time is no longer a straight line. It is a *curve*. At any particular time, the rate of change is often referred to as the 'instantaneous rate of change'. It is measured as '*the slope of the tangent to the curve at that particular time*' and is represented by the expression dy/dt. (The notation d/dt can be interpreted as 'instantaneous rate of change with respect to time'.) It is not easy to draw the tangent to a curve at a particular point. If the real experimental data consist of measurements at discrete points then it is first necessary to assume that these points are linked by a smooth curve and then to draw this curve. Again, this is not easy to accomplish although reasonable efforts can sometimes be achieved 'by eye'. A better approach is to use appropriate computer software. Even so, the best curve that can be computed will always be an approximation to the true curve and will also depend on the quality of the experimental data; for example in a chemical kinetic investigation on how well concentrations can be measured at specific times. The uncertainty in the value of the tangent that is computed at any point will reflect these factors.

STUDY NOTE

Do not miss out Exercise 1.1 below which, as already indicated, provides you with the opportunity to test your understanding of key ideas relating to rate of change. In addition it provides you with your first opportunity to use the *Kinetics Toolkit* on the CD-ROM. **It is important that you become familiar with the use of this software**.

EXERCISE 1.1

Two cars (A and B) are travelling along a dual carriageway. When they reach a certain speed, a stopwatch is started and the distance they travel is then monitored every 10 s over a period of one minute. The results obtained are summarized in Table 1.1 and plotted in Figures 1.5a and b, respectively. In each plot the best line, as judged by eye, that passes through all of the data points has been drawn.

Table 1.1 Distance versus time data for cars A and B

time/s	distance/m	
	car A	car B
	0	0
10	179	190
20	357	402
30	536	637
40	716	894
50	894	1 173
60	1 073	1 475

(a) Determine *directly* from the plots in Figure 1.5 the speed of each car after 40 s and, in each case, try and identify the main uncertainty in the calculation.

(b) Use the *Kinetics Toolkit*, in conjunction with the data in Table 1.1, to make your *own* plots of distance versus time for each car. Using a suitable analysis for each plot, once again determine the speed of each car after 40 s.

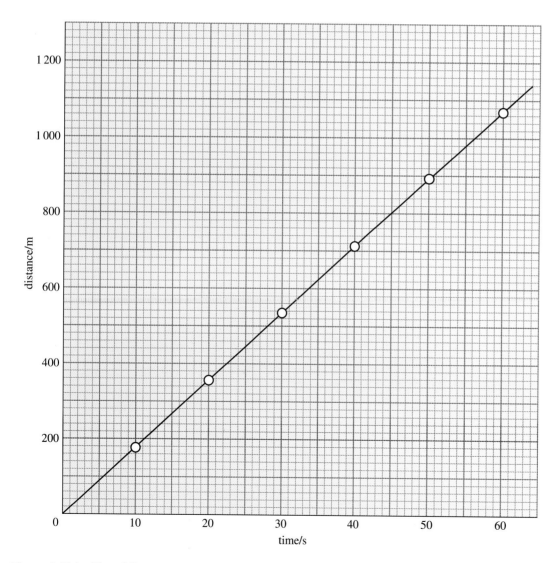

Figure 1.5(a) Plot of distance versus time for car A.

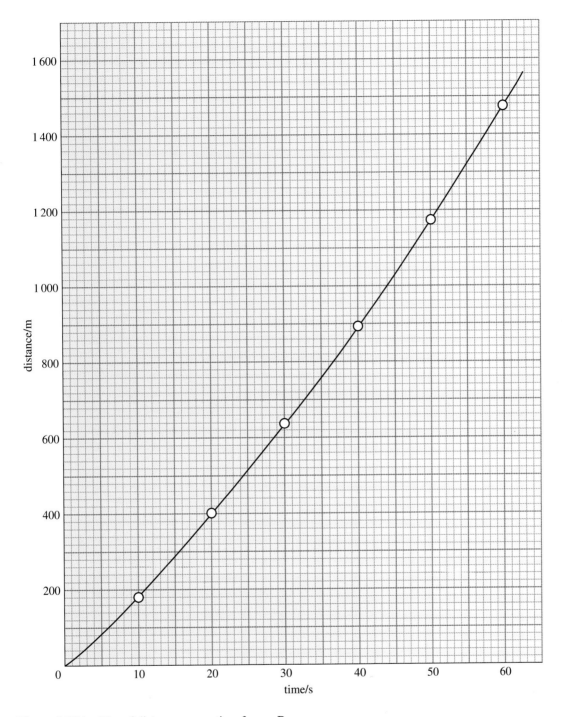

Figure 1.5(b) Plot of distance versus time for car B.

A CLOSER LOOK AT CHEMICAL REACTIONS

2

Around the beginning of the nineteenth century, the early chemists concentrated much of their effort on working out the proportions in which substances combine with one another and in developing a system of shorthand notation for representing chemical reactions. As a result, when we now think of the interaction of hydrogen and oxygen, for example, we tend to think automatically in terms of a *balanced* chemical equation

$$2H_2(g) + O_2(g) = 2H_2O(g) \qquad (2.1)$$

This equation serves to identify the species taking part and shows that for every two H_2 molecules and one O_2 molecule that react, two molecules of water are formed. This information concerning the *relative* amounts of reactants and products is known as the **stoichiometry** of the reaction. This term was introduced by the German chemist Jeremias Benjamin Richter as early as 1792 in order to denote the relative amounts in which acids and bases neutralize each other; it is now used in a more general way.

Important as it may be, knowing the stoichiometry of a reaction still leaves open a number of fundamental questions:

*   Does the reaction occur in a single step, as might be implied by a balanced chemical equation such as Equation 2.1, or does it involve a number of sequential steps?

*   In any step, are bonds broken, or made, or both? Furthermore, which bonds are involved?

*   In what way do changes in the relative positions of the various atoms, as reflected in the stereochemistry of the final products, come about?

*   What energy changes are involved in the reaction?

Answering these questions, particularly in the case of substitution and elimination reactions in organic chemistry, will be the main aim of a large part of this book. As you will see the key information that is required is embodied in the **reaction mechanism** for a given reaction. Broadly speaking, this refers to a *molecular description* of how the reactants are converted into products during the reaction. It is important at the outset to emphasize that a reaction mechanism is only as good as the information on which it is based. Essentially, it is a proposal of how a reaction is *thought* to proceed and its plausibility is always subject to testing by new experiments. For many mechanisms, we can be reasonably confident that they are correct, but we can never be completely certain.

A powerful means of gaining information about the mechanism of a chemical reaction is via experimental investigations of the way in which the reaction rate varies, for example, with the concentrations of species in the reaction mixture, or with temperature. There is thus a strong link between, on the one hand, experimental study and, on the other, the development of models at the molecular level. In the sections that follow we shall look in some depth at the principles that underlie experimental chemical kinetics before moving on to discuss reaction mechanism.

However, it is useful to establish a few general features relating to reaction mechanisms at this stage. In particular we look for features that relate to the steps involved and the energy changes that accompany them.

2.1 Individual steps

If we consider the reaction between bromoethane (CH_3CH_2Br) and sodium hydroxide in a mixture of ethanol and water at 25 °C then the stoichiometry is represented by the following equation

$$CH_3CH_2Br(aq) + OH^-(aq) = CH_3CH_2OH(aq) + Br^-(aq) \qquad (2.2)$$

where we have represented the states of all reactants as aqueous (aq). It is well established (and more to the point, no evidence has been found to the contrary) that this reaction occurs in a *single step*. We refer to it as an **elementary reaction**. For Reaction 2.2, therefore, the balanced chemical equation does actually convey the essential one-step nature of the process. The reaction mechanism, although consisting of only one step, is written in a particular way

$$CH_3CH_2Br + OH^- \longrightarrow CH_3CH_2OH + Br^- \qquad (2.3)$$

The arrow sign ($\longrightarrow$) is used to indicate that the reaction is known (or postulated) to be elementary and, by convention, the states of the species involved are not included. (Arrow signs are also used in this course in a more general way, particularly for organic reactions, to indicate that one species is converted to another under a particular set of conditions. The context in which arrow signs are used, however, should always make their significance clear.)

The reaction between phenylchloromethane ($C_6H_5CH_2Cl$) and sodium hydroxide in water at 25 °C

$$C_6H_5CH_2Cl(aq) + OH^-(aq) = C_6H_5CH_2OH(aq) + Cl^-(aq) \qquad (2.4)$$

is of a similar type to that in Reaction 2.2. However, all of the available experimental evidence suggests that Reaction 2.4 does *not* occur in a single elementary step. The most likely mechanism involves two steps

$$C_6H_5CH_2Cl \longrightarrow [C_6H_5CH_2]^+ + Cl^- \qquad (2.5)$$
$$[C_6H_5CH_2]^+ + OH^- \longrightarrow C_6H_5CH_2OH \qquad (2.6)$$

A reaction such as this, because it proceeds via *more than one* elementary step, is known as a **composite reaction**. The corresponding mechanism, Reactions 2.5 and 2.6, is referred to as a **composite reaction mechanism**, or just a *composite mechanism*. In general, for any composite reaction, the number and nature of the steps in the mechanism *cannot* be deduced from the stoichiometry. This point is emphasized when we consider that the apparently simple reaction between hydrogen gas and oxygen gas to give water vapour (Reaction 2.1) is thought to involve a sequence of up to 40 elementary steps.

The species $[C_6H_5CH_2]^+$ in the mechanism represented by Reactions 2.5 and 2.6 is known as a **reaction intermediate**. (This particular species, referred to as a carbocation, has a trivalent carbon atom which normally takes the positive charge. Carbocations are discussed in more detail in Part 2 of this book.) All mechanisms with more than a single step will involve intermediate species and these will be formed in one step and consumed, in some way, in another step. It is worth noting, although without going into detail, that many intermediate species are extremely reactive and short-lived which often makes it very difficult to detect them in a reaction mixture.

What is the result of adding Equations 2.5 and 2.6 together?

The addition gives

$$C_6H_5CH_2Cl + [C_6H_5CH_2]^+ + OH^- \longrightarrow [C_6H_5CH_2]^+ + Cl^- + C_6H_5CH_2OH$$

Cancelling the reaction intermediate species from both sides of the equation gives

$$C_6H_5CH_2Cl + OH^- \longrightarrow C_6H_5CH_2OH + Cl^-$$

In other words, adding the two steps together gives the form of the balanced chemical equation.

In general, for most composite mechanisms the sum of the various steps should add up to give the overall balanced chemical equation. (An important exception is a radical chain mechanism; see *Further reading* for a reference to these types of mechanism.)

It is a matter of general experience, that chemical reactions are not instantaneous. Even explosions, although extremely rapid, require a finite time for completion. This resistance to change implies that at the *molecular level* individual steps in a mechanism require energy in order to take place. For a given step, the energy requirement will depend on the species involved.

A convenient way to depict the energy changes that occur during an elementary reaction is to draw, in a *schematic* manner, a so-called **energy profile**; an example is given in Figure 2.1. The vertical axis represents potential energy which has contributions from the energy stored within chemical bonds as well as that associated with the interactions between each species and its surroundings. The horizontal axis is the **reaction coordinate** and this represents the path the system takes in passing from reactants to products during the reaction event.

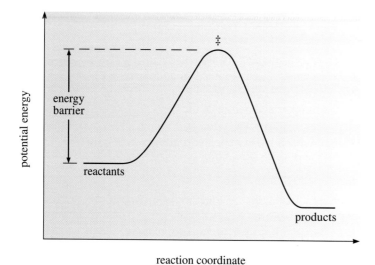

Figure 2.1 A schematic energy profile for a chemical reaction.

An energy profile such as that in Figure 2.1 can be interpreted in two distinct ways; either as representing the energy changes that occur when individual molecular species interact with one another in a single event, or as representing what happens on a macroscopic scale, in which case some form of average has to be taken over many billions of reactions. It is useful to consider the molecular level description first.

If we take the elementary reaction in Equation 2.3 as an example then from a molecular viewpoint, the energy profile shows the energy changes that occur when a single bromoethane molecule encounters, and reacts with, a single hydroxide ion in solution. As these species come closer and closer together they interact and, as a consequence, chemical bonds become distorted and the overall potential energy increases. At distances typical of chemical bond lengths, the reactant species become partially bonded together and new chemical bonds begin to form. At this point the potential energy reaches a maximum and any further distortion then favours the formation of product species and a corresponding fall in potential energy. It is, of course, possible to imagine that a bromoethane molecule and a hydroxide ion, particularly in the chaotic environment of the solution at the molecular level, will approach one another in a wide variety of ways. Each of these approaches will have its own energy profile.

The situation at the potential energy maximum is referred to as the **transition state** and it is often represented by the symbol ‡ (pronounced as 'double-dagger'). The molecular species that is present at this energy maximum is one in which old bonds are breaking and new ones are forming: it is called the **activated complex**. It is essential to recognize that this complex is a *transient* species and not a reaction intermediate. (It is worth noting that the terms 'activated complex' and 'transition state' are sometimes used incorrectly, in referring to both the transient species itself and the point of maximum potential energy.) Gaining information on what happens *within* the transition state is of fundamental interest.

● So far, we have characterized an elementary reaction as one that occurs in a single step. How would you further qualify this statement?

● For an elementary reaction we can specify that (i) it does *not* involve the formation of any reaction intermediate, and (ii) it passes through a *single* transition state.

It is clear in Figure 2.1 that there is an **energy barrier to reaction**. So, for example, for a bromoethane molecule to react with a hydroxide ion, energy must be supplied to overcome this barrier. The source for this energy is the kinetic energy of collision between the two species in solution; in crude terms the more violent the collision process, the more likely a reaction will occur.

● If you look at the elementary reaction in Equation 2.5, do you see a problem with this argument?

● The implication is that this is an elementary reaction involving a *single reactant molecule*. No other species appear to take part, which would seem to rule out the possibilities of collisions, and yet energy will certainly be required to break the C—Cl bond; the reactant molecule will not simply fall apart of its own volition.

The answer to the apparent anomaly is that energy is supplied by collisions with other $C_6H_5CH_2Cl$ molecules, or with solvent molecules. In this way the decomposition of $C_6H_5CH_2Cl$ can take place to give the reaction intermediate

$[C_6H_5CH_2]^+$ and Cl^-. In fact this general idea of 'other collisions' is part of a much more detailed theory first put forward in the doctoral thesis of J. A. Christiansen in 1921, but later attributed to a more senior worker, F. A. Lindemann, in 1922.

If we now turn to a macroscopic interpretation of the energy profile in Figure 2.1 then we can still retain the ideas of a transition state and an activated complex. The energy barrier to reaction is now a very complex average over many molecular events but, as we shall see later, it can still be related to a quantity that is measured experimentally. From a thermodynamic viewpoint, the energy difference between the products and reactants can be taken — to a good approximation — to be equal to the *enthalpy change for the elementary reaction.*

● Does the energy profile in Figure 2.1 represent an exothermic or endothermic change?

● The difference (measured as 'products minus reactants') in potential energy is negative. Thus the enthalpy change will be negative and the elementary reaction is exothermic.

QUESTION 2.1

Given that the elementary reaction in Equation 2.5 is endothermic, sketch and label an energy profile. What can you deduce about the magnitude of the energy barrier to reaction from this energy profile?

It is also possible to draw a schematic energy profile for a composite reaction; this will consist of the energy profiles for the individual elementary steps. For a two-step mechanism, such as that represented by Reactions 2.5 and 2.6, a possible energy profile would be as shown in Figure 2.2. Note that the horizontal axis is still labelled 'reaction coordinate', although this should *not* be taken to imply that the second step occurs immediately on completion of the first. The intermediate carbocation may undergo many, many collisions with various species before finally experiencing a successful collision with an OH^- ion as represented by Equation 2.6.

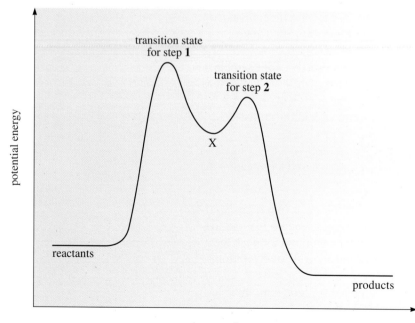

Figure 2.2
A schematic energy profile for a two-step composite mechanism with the steps labelled **1** and **2**.

● Is the overall reaction in Figure 2.2 exothermic or endothermic?

● From a thermodynamic viewpoint, it is *only* the initial and final states of a composite reaction that need to be considered. Overall, the reaction is exothermic.

● What is the significance of the point marked X in Figure 2.2?

● It is a local minimum that corresponds to the formation of the reaction intermediate; that is, in the case of the mechanism represented by Reactions 2.5 and 2.6, the species $[C_6H_5CH_2]^+$.

2.2 Summary of Section 2

1   Information concerning the relative amounts of reactants and products taking part in a chemical reaction is known as the stoichiometry of the reaction.

2   In general terms, a reaction mechanism provides a molecular description of how reactants are thought to be converted into products during a chemical reaction.

3   An elementary reaction is one that takes place in a single step, does not involve the formation of any intermediate species, and which passes through a single transition state.

4   A chemical reaction that proceeds by a series of elementary steps is known as a composite reaction and the corresponding mechanism is referred to as a composite reaction mechanism, or just composite mechanism.

5   All reaction mechanisms with more than one step will involve intermediate species. These are formed in one step and consumed, in some way, in another.

6   For composite mechanisms (except for radical chain mechanisms) the sum of the various steps gives the overall balanced chemical equation.

7   Any elementary reaction can be represented by a schematic energy profile which can be interpreted at the molecular level or on a macroscopic scale.

8   The transition state ($\ddagger$) lies at the top of the energy barrier to reaction; the species at the top of this barrier is transient and is called the activated complex.

9   The energy barrier to reaction for an endothermic elementary reaction must be at least as large as the corresponding enthalpy change.

RATE IN CHEMICAL KINETICS

3

In Section 1.1 we discussed rate in a general way, particularly in cases where time was involved. We now turn our attention to rate in chemical kinetics and, in particular, consider how to define the rate of a chemical reaction. Ideally, this quantity should have the *same, positive* value, regardless of whether it is defined in terms of a reactant or product species.

One of the main examples we shall use in this section is a reaction involving hypochlorite ions (ClO^-) and bromide ions in aqueous solution[*] at room temperature

$$ClO^-(aq) + Br^-(aq) = BrO^-(aq) + Cl^-(aq) \tag{3.1}$$

where BrO^- is the hypobromite ion. The stoichiometry is such that one mole of each reactant is converted into one mole of each product; in shorthand notation we refer to this as '1:1 stoichiometry'. It is important to emphasize that the determination of stoichiometry is an essential preliminary step in any kinetic study.

3.1 Kinetic reaction profiles

A kinetic study involves following a reaction as a function of *time*. This can be achieved by using a suitable analytical technique to measure the concentrations of reactants, or products, or both, at different times during the progress of the reaction. To avoid any changes in reaction rate due to temperature changes, it is essential that measurements are made under *isothermal*, that is constant temperature, conditions. A typical set of results obtained for Reaction 3.1 at a constant temperature of 25 °C is shown in Figure 3.1. This type of plot is called a **kinetic reaction profile**. This same term would also be used to describe a plot showing measurements for just a single reactant or product.

As is to be expected, Figure 3.1 shows that as the reaction proceeds, the concentrations of the two reactants decrease and the concentrations of the two products increase. In fact, the concentrations of the two products, BrO^- and Cl^-, change in exactly the same way. It should also be clear from Figure 3.1, that the *initial* conditions of the experiment were selected so that the initial concentration of ClO^- was greater than that of Br^-; we can say that ClO^- was in *excess*. In more concise terms, $[ClO^-]_0 > [Br^-]_0$ where the subscript zero has been used to indicate *initial* concentration. Experimentally, these initial concentrations were $[ClO^-]_0 = 3.230 \times 10^{-3}\ mol\ dm^{-3}$ and $[Br^-]_0 = 2.508 \times 10^{-3}\ mol\ dm^{-3}$.

---

[*] The solution must be in the pH range 10 to 14 to ensure that no other reactions take place; in particular to avoid chlorate (ClO_3^-) and bromate (BrO_3^-) ion formation.

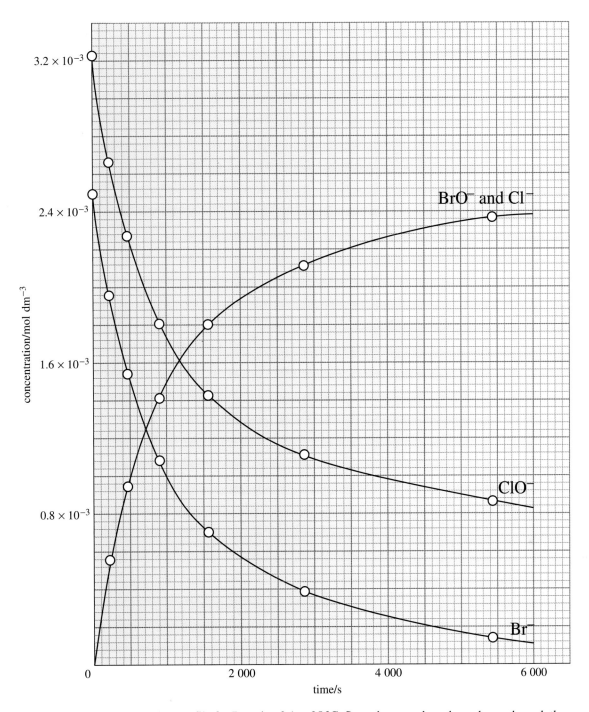

Figure 3.1 A kinetic reaction profile for Reaction 3.1 at 25 °C. Smooth curves have been drawn through the experimental data points. The behaviour for BrO⁻ and Cl⁻ is represented by a single curve.

BOX 3.1 Concentration units

Concentration is often expressed in moles per litre or $mol\,l^{-1}$.

Figure 3.2 shows a cube of dimension 10 cm. The volume of this cube is $10\,cm \times 10\,cm \times 10\,cm = 1\,000\,cm^3$, and this is equal to 1 litre. Alternatively, we can also recognize that 10 cm = 1 dm, where dm stands for decimetre. Thus the volume of the cube is also equal to $1\,dm \times 1\,dm \times 1\,dm = 1\,dm^3$.

Thus, units of moles per litre can also be expressed as moles per dm^3, or $mol\,dm^{-3}$.

Both $mol\,l^{-1}$ and $mol\,dm^{-3}$ are widely used as concentration units. However, the litre is not part of the SI system of units. $Mol\,dm^{-3}$ has a more formal status and is the concentration unit we use here.

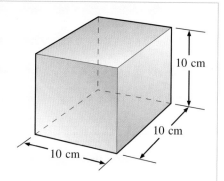

Figure 3.2
A cube of dimension 10 cm.

● By how much have the concentrations of ClO^-, Br^-, BrO^- and Cl^- changed after 2 000 s of reaction?

● The changes in concentration for ClO^- and Br^- correspond, in each case, to a decrease of about $1.95 \times 10^{-3}\,mol\,dm^{-3}$ compared to their initial concentrations. For BrO^- and Cl^- there is an increase in concentration from zero to about $1.95 \times 10^{-3}\,mol\,dm^{-3}$.

Thus, after 2 000 s of reaction the magnitudes of the changes in the concentrations of reactants and products in Reaction 3.1 are the same, although there is a decrease for reactants and an increase for products. In fact, this type of result would have been obtained irrespective of the time period selected. This means that the stoichiometry of Reaction 3.1 applies throughout the *whole* course of reaction; that is it has **time-independent stoichiometry**.

It might be tempting to conclude that intermediates are not present for a reaction that has time-independent stoichiometry. However, this is not the case. Time-independent stoichiometry simply means that, *within the accuracy of the chemical analysis used*, intermediates cannot be detected and so they do not affect the stoichiometric relationship between reactants and products. In fact, Reaction 3.1 is thought to be composite with a three-step mechanism in which case intermediates *must* be involved.

As a final point, it can be seen in Figure 3.1 that the concentration of Br^-, which is the reactant not in excess, appears to be progressing towards zero. In fact, after 10 hours in an extended experiment it was found that $[Br^-] = 0.007 \times 10^{-3}\,mol\,dm^{-3}$. Thus, nearly all of the Br^- has been consumed to give product. From a *kinetic viewpoint*, it is reasonable to say in these circumstances that the reaction has gone to **completion**, that is, had the reactants been initially present with equal concentrations, they would both have been virtually completed converted into products since there is greater than 99% reaction.

QUESTION 3.1

How would you summarize (in a single sentence) the features we have *determined*, so far, for Reaction 3.1?

3.2 Rate of change of concentration of a reactant or product with time

The concentration-versus-time profiles for each of the reactants and products in Figure 3.1 are curved. This means that the rate of change of concentration with time for each of these species is not constant; in each case it will vary continuously as the reaction progresses.

● How would you determine the rate of change of concentration with time for BrO^- at 1 500 s?

● The rate will be equal to the slope of the tangent drawn to the curve at 1 500 s. This will measure the *instantaneous* rate of change and will be represented by $d[BrO^-]/dt$. (If you are uncertain about any aspect of this answer you should look again at Section 1.1.)

Figure 3.3 plots a kinetic reaction profile for BrO^- and shows the tangent drawn to the curve at 1 500 s.

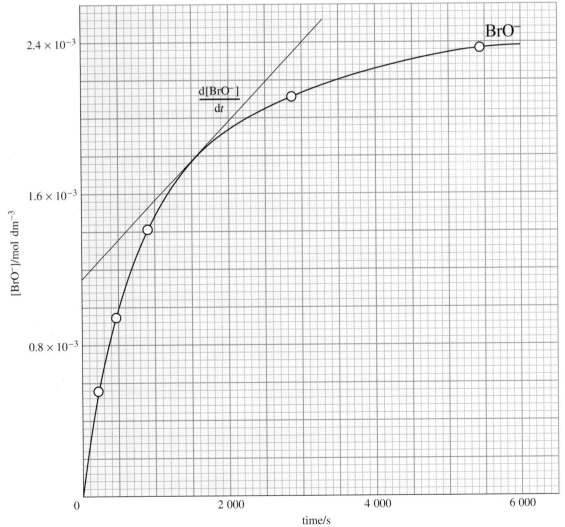

Figure 3.3
A kinetic reaction profile for BrO^- measured for Reaction 3.1 at 25 °C. A smooth curve is drawn through the experimental data points and the tangent is drawn at 1 500 s.

● If the coordinates for two points on the tangent are ($t = 0$ s, $[BrO^-] = 1.14 \times 10^{-3}$ mol dm^{-3}) and ($t = 4\,000$ s, $[BrO^-] = 2.82 \times 10^{-3}$ mol dm^{-3}), what is the value of $d[BrO^-]/dt$ at $1\,500$ s?

● The slope of the tangent is calculated as follows

$$\text{slope} = \frac{2.82 \times 10^{-3} \text{ mol dm}^{-3} - 1.14 \times 10^{-3} \text{ mol dm}^{-3}}{4\,000 \text{ s} - 0 \text{ s}}$$

$$= \frac{1.68 \times 10^{-3} \text{ mol dm}^{-3}}{4\,000 \text{ s}}$$

$$= 4.20 \times 10^{-7} \text{ mol dm}^{-3} \text{ s}^{-1}$$

This value, which is an estimate that depends on how well the tangent has been drawn, is the instantaneous rate of change of concentration of BrO$^-$ with time, $d[BrO^-]/dt$, at $1\,500$ s. Notice that the units for this quantity have been derived by including appropriate units at *all* stages in the calculation.

It becomes cumbersome to keep using the qualifying term, 'instantaneous'. From now on when we discuss, or calculate, a rate of change of concentration with time we shall always understand it to mean an instantaneous rate at a specific time.

Clearly, $d[BrO^-]/dt$ and $d[Cl^-]/dt$ will be equal in value since the kinetic reaction profiles for BrO$^-$ and Cl$^-$ are identical. They are also both positive quantities since they represent the *formation* of product species. Thus, if we represent the rate of Reaction 3.1 at any time by the symbol J, then one possible definition would be

$$J = \frac{d[BrO^-]}{dt} = \frac{d[Cl^-]}{dt} \tag{3.2}$$

However, we could equally well have considered reaction profiles for ClO$^-$ and Br$^-$ and determined $d[ClO^-]/dt$ and $d[Br^-]/dt$. Can these quantities also be used to define the rate of Reaction 3.1?

QUESTION 3.2

This question uses the *Kinetics Toolkit*. The experimental data that were used for plotting Figure 3.1, the kinetic reaction profile for Reaction 3.1, is given in Table 3.1. It is presented in a form suitable for *direct* entry into the graph-plotting software, although you should note that the 'E' format must be used for inputting powers of ten. For example, 3.230×10^{-3} is input as 3.230E−3. (*When you have input your data, you should store it in an appropriately named file.*)

(a) Determine values of $d[ClO^-]/dt$ and $d[Br^-]/dt$ at $1\,500$ s. (You may also wish to check the value of $d[BrO^-]/dt$.)

(b) Determine values of $d[ClO^-]/dt$, $d[Br^-]/dt$, $d[BrO^-]/dt$ and $d[Cl^-]/dt$ at $3\,000$ s.

Table 3.1 Data used for plotting the kinetic reaction profile in Figure 3.1

time	$[ClO^-]$	$[Br^-]$	$[BrO^-]$ or $[Cl^-]$
s	mol dm^{-3}	mol dm^{-3}	mol dm^{-3}
0	3.230×10^{-3}	2.508×10^{-3}	0
220	2.670×10^{-3}	1.948×10^{-3}	0.560×10^{-3}
460	2.277×10^{-3}	1.555×10^{-3}	0.953×10^{-3}
900	1.810×10^{-3}	1.088×10^{-3}	1.420×10^{-3}
1 560	1.430×10^{-3}	0.708×10^{-3}	1.800×10^{-3}
2 860	1.113×10^{-3}	0.391×10^{-3}	2.117×10^{-3}
5 435	0.863×10^{-3}	0.141×10^{-3}	2.367×10^{-3}

The results from Question 3.2 can be summarized as follows. The rates of change of concentration of ClO^- and Br^- at any time in the reaction (that is, $d[ClO^-]/dt$ and $d[Br^-]/dt$) are

- negative, because they represent the *consumption* of reactant species
- equal to one another, because according to the stoichiometry if a ClO^- ion reacts then so *must* a Br^- ion
- equal in magnitude, but opposite in sign, to $d[BrO^-]/dt$ and $d[Cl^-]/dt$, because according to the stoichiometry, the reaction of a ClO^- ion with a Br^- ion *must* produce one each of the two product ions.

It is important to note that these points will hold no matter which reactant species is in excess, and irrespective of the amount of the excess.

○ If at 1 500 s, $d[ClO^-]/dt = -4.18 \times 10^{-7}$ mol dm^{-3} s^{-1} (taken from the answer to Question 3.2) what will be the value of $-d[ClO^-]/dt$ at this time?

○ The value will be as follows

$$-\frac{d[ClO^-]}{dt} = -(-4.18 \times 10^{-7} \text{ mol dm}^{-3} \text{ s}^{-1})$$

$$= 4.18 \times 10^{-7} \text{ mol dm}^{-3} \text{ s}^{-1}$$

since taking the negative of a negative quantity gives a positive result.

The fact that $-d[ClO^-]/dt$ and $-d[Br^-]/dt$ are positive quantities puts us in a position to give a final definition of J for Reaction 3.1

$$J = -\frac{d[ClO^-]}{dt} = -\frac{d[Br^-]}{dt} = \frac{d[BrO^-]}{dt} = \frac{d[Cl^-]}{dt} \tag{3.3}$$

Defined in this way, J, *irrespective* of whether it is expressed in terms of a reactant or a product in Reaction 3.1, always has a single positive value at any time in the reaction.

One special case of the rate of reaction is that corresponding to the *start* of the reaction. This is referred to as the **initial rate of reaction** and is represented by J_0 ('J subscript zero'). Figure 3.4 plots a kinetic reaction profile for Cl^- and shows the tangent (labelled 'initial tangent') drawn to the curve so that the initial rate of change of concentration of Cl^- can be determined.

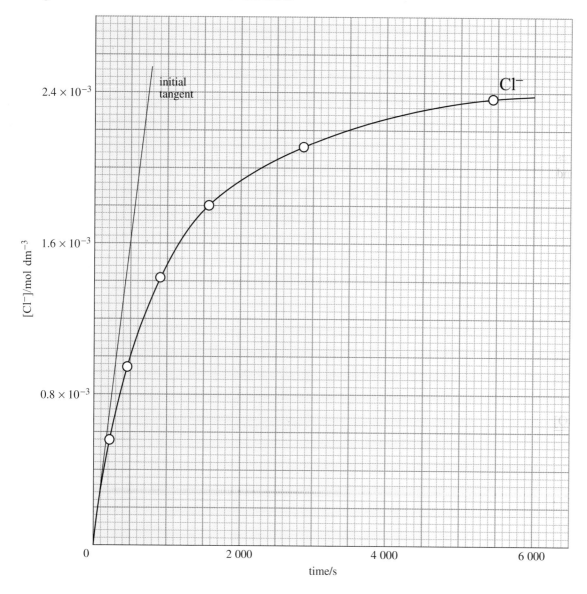

Figure 3.4 A kinetic reaction profile for Cl^- measured for Reaction 3.1 at 25 °C. A smooth curve is drawn through the experimental data points.

QUESTION 3.3

If the tangent in Figure 3.4 passes through a point with coordinates ($t = 500\,s$, $[Cl^-] = 1.60 \times 10^{-3}\,mol\,dm^{-3}$), what is the initial rate of reaction?

Experimentally there are two important factors that must be taken into account when measuring initial rates of reaction. Firstly, it is preferable to measure an initial rate by observing the appearance of a product rather than the disappearance of a reactant.

This is because higher analytical accuracy is needed to measure the relatively small changes in the initially high concentration of a reactant. Secondly, it is essential to make measurements in the very early stages, say the first 5%, of a reaction in order to obtain accurate values of the initial rate. For example, if you use the *Kinetics Toolkit* to determine the initial rate of Reaction 3.1 from the data in Table 3.1 then you will find $J_0 = 2.75 \times 10^{-6} \, \text{mol dm}^{-3} \, \text{s}^{-1}$. This value is significantly different from that calculated (in Question 3.3) from the initial tangent in Figure 3.4 which was drawn using information based on just the first 20 s of reaction.

3.3 A general definition of the rate of a chemical reaction

The discussion that resulted in Equation 3.3 can be applied to any chemical reaction that has time-independent stoichiometry. For example, nitrogen dioxide (NO_2) decomposes in the gas phase at temperatures in the region of 300 °C to give nitric oxide (NO) and oxygen

$$2NO_2(g) = 2NO(g) + O_2(g) \tag{3.4}$$

If the progress of this gas-phase reaction is monitored in a closed reaction vessel then concentrations can simply be expressed in terms of mol dm^{-3}.

At any time in the decomposition, the rate of *decrease* in the concentration of NO_2 will be directly related to the rates of *increase* in the concentrations of NO and O_2, respectively.

○ What is the relationship between these quantities?

○ If we consider $-\text{d}[NO_2]/\text{d}t$, which is a positive quantity, then

$$-\frac{\text{d}[NO_2]}{\text{d}t} = \frac{\text{d}[NO]}{\text{d}t} = 2\frac{\text{d}[O_2]}{\text{d}t}$$

These relationships are consistent with the fact that according to the stoichiometry, the rate of increase in the concentration of O_2 is only equal to one-half of that for NO.

The answer to the above question could equally well have been written in a fractional form, that is

$$-\frac{1}{2}\frac{\text{d}[NO_2]}{\text{d}t} = \frac{1}{2}\frac{\text{d}[NO]}{\text{d}t} = \frac{\text{d}[O_2]}{\text{d}t} \tag{3.5}$$

It is this form that is conventionally used to define the rate of reaction and so

$$J = -\frac{1}{2}\frac{\text{d}[NO_2]}{\text{d}t} = \frac{1}{2}\frac{\text{d}[NO]}{\text{d}t} = \frac{\text{d}[O_2]}{\text{d}t} \tag{3.6}$$

It is common practice in chemical kinetics (as well as in chemical thermodynamics) to use a chemical reaction written in an 'alphabetical' form to help to express a definition in a general way. Thus, we could write a chemical reaction with *known* stoichiometry as

$$a\text{A} + b\text{B} + \ldots = p\text{P} + q\text{Q} + \ldots \tag{3.7}$$

where A, B and so on, represent reactants and P, Q and so on, represent products.

STUDY NOTE

If you would like to look at this definition in more detail then information for plotting a kinetic reaction profile for Reaction 3.4 is given in Exercise 3.1 at the end of this section.

The numbers a, b … and p, q … ensure that the equation is balanced and so are known as *balancing coefficients*. In practice they are usually chosen to have their *smallest possible integer* values, and they must be *positive*. Writing a chemical reaction in this way allows a quantity called the **stoichiometric number** to be introduced. It is given the symbol v_Y (v is the Greek letter 'nu' and v_Y is pronounced 'nu Y') where the subscript Y represents a given species (reactant or product) in the reaction. The stoichiometric number is then defined so that for

reactant A, $v_A = -a$

reactant B, $v_B = -b$

product P, $v_P = +p$

product Q, $v_Q = +q$

Hence, the stoichiometric number for a reactant is always negative, and that for a product is always positive.

⬤ What are the stoichiometric numbers of NO_2, NO and O_2 in Reaction 3.4?

⬤ According to the definition given above

$$v_{NO_2} = -2 , \ v_{NO} = +2 \ \text{and} \ v_{O_2} = +1 .$$

In terms of Equation 3.7, we can now make a general definition for the **rate of a chemical reaction**

$$J = \frac{1}{v_A} \frac{d[A]}{dt} = \frac{1}{v_B} \frac{d[B]}{dt} = \frac{1}{v_P} \frac{d[P]}{dt} = \frac{1}{v_Q} \frac{d[Q]}{dt} \qquad (3.8)$$

⬤ Is the term $\dfrac{1}{v_A} \dfrac{d[A]}{dt}$ positive or negative in value?

⬤ It is positive. Since A is a reactant $d[A]/dt$ is negative. But the stoichiometric number for a reactant is defined to be negative. So, dividing $d[A]/dt$ by v_A is equivalent to dividing one negative quantity by another, and this gives a positive result.

It is important to remember, however, that this definition only holds for a reaction with time-independent stoichiometry. If, for example, intermediates build up to measurable quantities during the course of a reaction then there are no simple relationships between the rates of change of concentrations of reactants and products.

Equation 3.8 can be written in a more concise form if a reactant or a product in a reaction is simply represented by Y (as we have already done in our discussion of the stoichiometric number). It then follows that

$$J = \frac{1}{v_Y} \frac{d[Y]}{dt} \qquad (3.9)$$

Strictly, this definition assumes *constant volume* conditions during the course of a reaction. For solution reactions this is a reasonable approximation. It is also valid for gas-phase reactions carried out in sealed containers.

QUESTION 3.4

For the following two reactions express the rate of reaction, J, in terms of the rate of change of concentration of each reactant and each product. (Assume that both reactions have time-independent stoichiometry.)

(a) $2H_2(g) + 2NO(g) = 2H_2O(g) + N_2(g)$

(b) $S_2O_8^{2-}(aq) + 3I^-(aq) = 2SO_4^{2-}(aq) + I_3^-(aq)$

3.4 Summary of Section 3

1   The determination of reaction stoichiometry is a very important preliminary activity in any kinetic investigation.

2   A kinetic reaction profile is a plot of the concentrations of reactants or products in a reaction, individually or combined, as a function of time under isothermal conditions.

3   If the same stoichiometry for a reaction applies throughout the whole course of a reaction then the reaction is said to have time-independent stoichiometry.

4   If a single reactant, or a reactant that is not in excess, is almost totally consumed in a reaction then from a kinetic viewpoint the reaction is said to have gone to completion.

5   The instantaneous rate of change of the concentration of a reactant, or a product, at a particular instant in a chemical reaction is equal to the slope of the tangent drawn to the kinetic reaction profile at that time.

6   For a reaction of the form

$$aA + bB = pP + qQ$$

where the numbers a, b ..., p, q ..., have their smallest possible, positive, integer values, then the stoichiometric numbers are defined to be $v_A = -a$, $v_B = -b$, $v_P = +p$ and $v_Q = +q$. The stoichiometric number for a reactant is always negative, that for a product is always positive.

7   The rate of a chemical reaction (strictly at constant volume) for a reaction with time-independent stoichiometry is defined by

$$J = \frac{1}{v_Y} \frac{d[Y]}{dt}$$

where Y represents either a reactant or a product. This definition ensures that the rate of a chemical reaction is always a single, positive quantity, irrespective of whether it is defined in terms of a reactant or a product species.

8   The rate of reaction at the start of reaction is referred to as the initial rate of reaction, J_0.

QUESTION 3.5

The information in Table 3.1 is presented in a form suitable for direct entry into the graph plotting software in the *Kinetics Toolkit*. Often with such data, it is more usual to present it so that powers of ten are incorporated into the column headings. So, for example, the column of data for [ClO$^-$] would have *numbers* running from 3.230 to 0.863. In this case, what form would the column heading take?

EXERCISE 3.1

Data for plotting a kinetic reaction profile for the gas-phase decomposition of NO_2 (Reaction 3.4) at 300 °C are given in Table 3.2. It is presented in a form suitable for direct entry into the graph plotting software in the *Kinetics Toolkit*. (*When you have entered the data you should store them in a file for future use.*)

(a) Determine values of $d[NO_2]/dt$, $d[NO]/dt$ and $d[O_2]/dt$ at 500 s. Hence using any of these values, determine the rate of reaction, J, at 500 s.

(b) In addition, determine the rates of reaction at 1 000 s and 1 500 s.

Table 3.2 Data for plotting a kinetic reaction profile for the gas-phase decomposition of NO_2 (Reaction 3.4) at 300 °C

time $\dfrac{}{s}$	$\dfrac{[NO_2]}{mol\ dm^{-3}}$	$\dfrac{[NO]}{mol\ dm^{-3}}$	$\dfrac{[O_2]}{mol\ dm^{-3}}$
0	4.00×10^{-3}	0	0
100	2.83×10^{-3}	1.18×10^{-3}	0.59×10^{-3}
240	2.00×10^{-3}	2.00×10^{-3}	1.00×10^{-3}
320	1.72×10^{-3}	2.28×10^{-3}	1.14×10^{-3}
500	1.30×10^{-3}	2.70×10^{-3}	1.35×10^{-3}
780	0.94×10^{-3}	3.06×10^{-3}	1.53×10^{-3}
1 000	0.78×10^{-3}	3.22×10^{-3}	1.61×10^{-3}
1 500	0.55×10^{-3}	3.46×10^{-3}	1.73×10^{-3}
2 000	0.43×10^{-3}	3.56×10^{-3}	1.78×10^{-3}

FACTORS DETERMINING THE RATE OF A CHEMICAL REACTION

4

4.1 A simple collision model

The essential theoretical picture in chemical kinetics is that for a step in a reaction mechanism to occur, two things must happen:

- reactant species involved in the step must collide with one another, and
- colliding particles must have sufficient energy to overcome the energy barrier separating reactants from products (Section 2.1).

The essential properties of solutions or gases tell us that the constituent particles are always in constant, random motion. We can envisage, therefore, that collisions occur continuously and this suggests that the more frequently they do so between *reactant* species, then the faster the consequent reaction. It is useful to look at this idea in a little more detail.

As an example we can consider an elementary reaction between two different species A and B (which could be molecules, 'fragments of molecules', atoms or ions) in the gas phase

$$A + B \longrightarrow products \tag{4.1}$$

The number of collisions between species A and species B that occur in a fixed volume in unit time (say, 1 s) is a measure of the *collision rate* between A and B. This rate will depend upon the concentration of *both* species. For example, doubling the concentration of B means that the number of targets for individual A species in a given volume is increased by a factor of two; hence, the rate at which A species collide with B species is doubled. A similar argument holds for increasing the concentration of A. Thus, overall, the collision rate between the A and B species is directly proportional to their *concentrations multiplied together*, so that

$$\text{collision rate} \propto [A][B] \tag{4.2}$$

or $\quad$ collision rate $= c \times [A][B]$ $\hspace{3cm}$ (4.3)

where c is a constant of proportionality. In fact, the form of this constant can be calculated for any gas-phase elementary reaction using a theory of collisions in the gas phase that was first put forward in the 1920s.

If *every* collision between species A and B resulted in chemical transformation to products, then the rate of reaction (J) would be identical to the collision rate. For many elementary reactions, however, this is not the case.

- Can you suggest the reason for this?

- In general terms, as discussed in Section 2.1, there is an energy barrier to reaction for an elementary reaction. If the kinetic energy involved in a collision is insufficient to overcome this barrier then the colliding species simply move apart again.

Of all the collisions that occur between reactant species A and B then, only a fraction, f, will be successful. We can therefore write the rate of reaction as

$$J = f \times (\text{collision rate}) \tag{4.4}$$

or, using Equation 4.3

$$J = f \times c \times [A][B] \tag{4.5}$$

In the discussion so far we have implicitly assumed that the temperature is *fixed*. This being the case, the quantity $f \times c$ in Equation 4.5 can be replaced by a single constant, k_{theory}, so that

$$J = k_{theory}[A][B] \tag{4.6}$$

This equation is an example of a rate equation and, more explicitly, it is the **theoretical rate equation** for the elementary reaction described by Equation 4.1. The quantity k_{theory} is the **theoretical rate constant** for the elementary reaction; it has a value that is *independent* of the concentrations of reactants A and B.

⬤ If the units of J in Equation 4.6 are expressed as $mol\,dm^{-3}\,s^{-1}$, what are the units of k_{theory}?

⬤ The units of k_{theory} can be calculated from $J/[A][B]$ with the units of concentration expressed in $mol\,dm^{-3}$. So, the units are

$$(mol\,dm^{-3}\,s^{-1})/(mol\,dm^{-3})(mol\,dm^{-3})$$

This can be simplified

$$\frac{mol\,dm^{-3}\,s^{-1}}{(mol\,dm^{-3})(mol\,dm^{-3})} = \frac{s^{-1}}{mol\,dm^{-3}}$$
$$= mol^{-1}\,dm^3\,s^{-1}$$

So, the units of k_{theory} *in this particular case* are $mol^{-1}\,dm^3\,s^{-1}$. (Often this is written as $dm^3\,mol^{-1}\,s^{-1}$ so that the unit with the positive exponent comes first. This is the practice we shall adopt.)

The theoretical rate constant for Reaction 4.1, although called a 'constant', does depend on temperature. Increasing the temperature *increases*, in most circumstances, the magnitude of k_{theory}. So carrying out a reaction at a higher temperature, but with the same initial concentrations of A and B, will be expected to result in an increase in the rate of reaction. This behaviour can be understood in a qualitative way in terms of a simple collision model.

In any gas, at a particular instant, the particles will be moving about with a wide distribution of speeds. The form of this distribution depends on temperature and was worked out towards the end of the nineteenth century by the Scottish scientist James Clerk Maxwell. It is shown in a schematic form for a gas (consisting of molecules) at two different temperatures in Figure 4.1. It is worth noting that the area under each curve is the same and is a constant for a given sample since it represents the total number of molecules in that sample.

Increasing the temperature clearly results in an increase in the number of more rapidly moving molecules, at the expense of the numbers moving more slowly, and the distribution becomes flatter and wider. Furthermore, the peak of the distribution, which corresponds to the most probable speed, moves to a higher value.

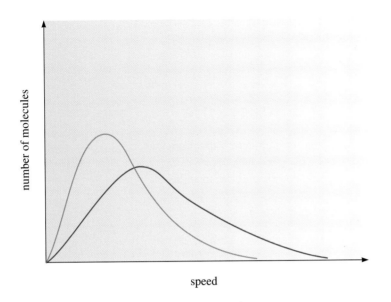

Figure 4.1
The schematic form of the distribution of molecular speeds in a gas at two different temperatures. The blue line corresponds to the lower temperature, and the red line to the higher temperature.

For an elementary reaction, such as that in Equation 4.1, raising the temperature has two distinct consequences.

- There is an increase in the *fraction* of rapidly moving species for both reactants. In turn this means that the fraction of collisions (represented by f in Equation 4.4) with a kinetic energy sufficient to overcome the energy barrier to reaction also increases. The effect on the rate of reaction can be quite significant.

- There is a general increase in speed for both reactant species and this results in an increased collision rate; in other words the constant c in Equation 4.3 is temperature dependent. This effect, however, is relatively small and for increases of temperature over a range of, say, 100 °C, it would be difficult to detect experimentally.

What would be the effect of an increase in temperature for an elementary reaction of the form of Equation 4.1 with an energy barrier to reaction that was very close to zero?

In these circumstances most collisions will lead to chemical reaction; in other words, the fraction of successful collisions will be effectively constant and independent of temperature. Thus, *experimentally*, it would be difficult to detect any change in the rate of reaction with increasing temperature.

The above discussion has been based on simple qualitative ideas about how an elementary reaction may occur. The way to test this picture, of course, is to see if rates of reaction measured experimentally, using different concentrations of each reactant and at different temperatures, show the same predicted behaviour. For this purpose the **experimental rate equations** for a few elementary reactions involving two reactant species are given in Table 4.1. In each case the **experimental rate constant** is denoted by the symbol k_R. Comparison of the form of the experimental rate equations in Table 4.1 with Equation 4.6 makes it clear that there is a good agreement between theory and experiment. Other details also help to confirm this conclusion. For example, the experimental rate constant for the reaction between potassium atoms and Br_2 molecules is found to be independent of temperature, suggesting that the energy barrier to reaction is effectively zero. By contrast, the rate constants for the other two reactions (in Table 4.1) are markedly temperature dependent.

Table 4.1 Experimental studies of elementary reactions

Reaction	Experimental rate equation
$K(g) + Br_2(g) = KBr(g) + Br(g)$	$J = k_R[K][Br_2]$
$Cl(g) + H_2(g) = HCl(g) + H(g)$	$J = k_R[Cl][H_2]$
$NO(g) + Cl_2(g) = NOCl(g) + Cl(g)$	$J = k_R[NO][Cl_2]$

The use of a simple collision model to predict the behaviour of elementary reactions involving two reactant species is instructive but nonetheless limited in scope. To extend such a model to chemical reactions in general would be difficult because the vast majority of these are composite. *To make progress in understanding the rates of chemical reactions it is necessary to adopt an experimental approach.*

4.2 An experimental approach

Experimental investigation, under isothermal conditions, for a wide range of chemical reactions which can be represented in a general form as

$$aA + bB + cC + \ldots = pP + qQ + rR + \ldots \qquad (4.7)$$

has shown in very many cases, both in the gas and solution phase, that the experimental rate equation takes the form

$$J = k_R[A]^\alpha[B]^\beta[C]^\gamma\ldots \qquad (4.8)$$

where k_R is the experimental rate constant. Two key points about this relationship are

- one concentration term appears for each reactant
- each concentration term is raised to a particular power: α, β, γ, and so on.

It is important to emphasize that the relationship is *empirical* in that it represents a generalization of the simplest mathematical way of representing the experimental rate equations for each of the reactions studied. The relationship is based simply on the results of observation and experiment.

The powers to which the concentration terms are raised in Equation 4.8 are known as **partial orders of reaction**. Thus α is the partial order of reaction with respect to reactant A, β is the partial order of reaction with respect to reactant B, γ is the partial order of reaction with respect to reactant C, and so on. The **overall order of reaction** (n) is defined by the sum of the partial orders

$$n = \alpha + \beta + \gamma + \ldots \qquad (4.9)$$

It is often, but not always, the case that the partial orders of reaction turn out to be small integers. If the partial order for a reactant is either 1 or 2, then the reaction is referred to as being *first-order* or *second-order* in that particular reactant. The most frequently observed values of overall order n are also 1 and 2 and the corresponding reactions are then referred to as being, respectively, first- and second-order processes. An overall order of reaction can *only* defined for a reaction that has an experimental rate equation corresponding to the general form given in Equation 4.8.

A few selected examples of reactions with experimental rate equations of the form in Equation 4.8 are given in Table 4.2.

Table 4.2 Experimental rate equations determined under isothermal conditions

Reaction	Experimental rate equation
(a) $S_2O_8^{2-}(aq) + 3I^-(aq) = 2SO_4^{2-}(aq) + I_3^-(aq)$	$J = k_R[S_2O_8^{2-}][I^-]$
(b) $3ClO^-(aq) = ClO_3^-(aq) + 2Cl^-(aq)$	$J = k_R[ClO^-]^2$
(c) $BrO_3^-(aq) + 5Br^-(aq) + 6H^+(aq) = 3Br_2(aq) + 3H_2O(l)$	$J = k_R[BrO_3^-][Br^-][H^+]^2$
(d) $(CH_3)_3CBr(aq) + OH^-(aq) = (CH_3)_3COH(aq) + Br^-(aq)$	$J = k_R[(CH_3)_3CBr]$
(e) $CO(g) + Cl_2(g) = COCl_2(g)$	$J = k_R[CO][Cl_2]^{1.5}$

● Reaction (a) in Table 4.2 is that between persulfate ion ($S_2O_8^{2-}$) and iodide ion (I^-) in aqueous solution. What are the partial orders with respect to these reactants and the overall order of reaction?

● The experimental rate equation is

$$J = k_R[S_2O_8^{2-}][I^-]$$

In this equation the concentration of $S_2O_8^{2-}$ is raised to the power of 1; the partial order with respect to $S_2O_8^{2-}$ is therefore 1 and the reaction is first-order in $S_2O_8^{2-}$. Similarly, the partial order with respect to I^- is 1 and the reaction is therefore also first-order in this reactant. The overall order is $n = 1 + 1 = 2$, that is second-order overall.

It is very important to recognize in this example that there is *no* simple link between the stoichiometry of the reaction and the form of the experimental rate equation. Trying to equate the partial orders of reaction for $S_2O_8^{2-}$ and I^- to their balancing coefficients in the chemical equation would be similar to trying to relate apples to pears. *It cannot be overemphasized that partial orders of reaction can be determined only from experimental measurements of the kinetics of a process.* In the case of $S_2O_8^{2-}$ it turns out by coincidence, and no more than this, that the partial order has the same value as the balancing coefficient. For I^-, the partial order and the balancing coefficient (equal to 3) are very different.

QUESTION 4.1

For reactions (b) and (c) in Table 4.2 what are the partial orders of reaction with respect to the individual reactants and the overall order of reaction in each case?

For reaction (d) in Table 4.2, the experimental rate equation does not depend on the concentration of one of the reactants in the chemical equation, that is the hydroxide ion, OH^-. In this case the reaction is said to be *zero-order* in OH^-.

● Can you suggest why the term 'zero-order' is used?

● If the experimental rate equation is written in the general form suggested by Equation 4.8, then it would be

$$J = k_R[(CH_3)_3CBr]^\alpha[OH^-]^\beta$$

where clearly $\alpha = 1$. Since the rate of reaction does not depend on $[OH^-]$ then the only possible value for β is zero; since $[OH^-]^0 = 1$. (Raising any quantity to the power zero gives a value of unity.) Thus, the reaction is zero-order in OH^- and a change in the concentration of this reactant does not affect the rate of reaction.

Reaction (e) in Table 4.2 demonstrates that a partial order of reaction may be *fractional*; the partial order with respect to Cl_2 is 1.5. This type of behaviour is often found for gas-phase reactions which have a particular type of mechanism.

The most important conclusion to be drawn from Table 4.2 is that:

> There is no systematic relationship between the stoichiometry of a reaction and the partial orders of reaction that are determined by experiment.

The only exception to this general conclusion is in the case of reactions which, according to all the available evidence, are elementary. This is discussed in more detail in Section 7. However, for now it can be noted, as demonstrated by the results in Table 4.1, that a simple collision theory can predict the form of the experimental rate equation for an elementary reaction involving two reactant species. For reactions which are not elementary, such as those in Table 4.2, no such theoretical approach is available. Indeed, if it were, then a large area of experimental chemical kinetics would never have come into existence.

> The rate equation for a chemical reaction, which provides information on the partial orders of reaction *and* the rate constant, has to be determined experimentally.

Partial orders of reaction are of more interest than the overall order. Essentially, the overall order of reaction provides a convenient means of *categorizing* reactions, but otherwise is of little importance.

In the next two sections we shall look in some detail at how experiments can be designed, and how the resulting data can be analysed, to obtain the form of a rate equation for a chemical reaction under a given set of experimental conditions. In subsequent sections we shall see that it is the values of partial orders of reaction, together with the value of the rate constant and the way in which it varies with temperature, that enable us to propose detailed mechanisms for reactions such as those in Table 4.2, among many others.

QUESTION 4.2

Reactions involving a single reactant, say A, can be written in the general form

aA = products

What would be the form of the experimental rate equation in the case that (a) the reaction was first-order overall or (b) the reaction was second-order overall? In each case what would be the units of the experimental rate constant, assuming that time is measured in seconds and concentration in $mol \, dm^{-3}$?

4.3 Summary of Section 4

1   The key idea underlying a simple collision model for an elementary reaction is that the reactant species must collide before any chemical transformation can take place.

2　For an elementary reaction involving two reactant species A and B, a simple collision model predicts a theoretical rate equation of the form

$$J = k_{theory}[A][B]$$

where k_{theory} is a theoretical rate constant. This equation is of the same form as that found in experimental studies of reactions that are thought to be elementary.

3　According to the simple collision model, the fraction of collisions with a kinetic energy sufficient to overcome the energy barrier to reaction increases with increasing temperature. This behaviour largely accounts for the temperature dependence of the theoretical rate constant.

4　For many chemical reactions, the experimental rate equation can be written in the form

$$J = k_R[A]^\alpha[B]^\beta[C]^\gamma \dots$$

where A, B, C, and so on, are reactant species. The quantities α, β, γ, and so on, are partial orders of reaction with respect to the individual reactant species. The quantity k_R is the experimental rate constant. The overall order of reaction is defined by

$$n = \alpha + \beta + \gamma + \dots$$

5　Partial orders of reaction are experimental quantities and often turn out to have small integer values. If the partial order is 1 for a particular reactant, then the reaction is first-order with respect to that reactant. If the partial order is 2 the reaction is second-order with respect to that reactant.

6　There is no systematic relationship between the stoichiometry of a reaction and the experimental partial orders of reaction.

DETERMINING EXPERIMENTAL RATE EQUATIONS AT A FIXED TEMPERATURE

5

5.1 Practical matters

As has already been mentioned in Section 3, a vital first step in any kinetic study is to determine the stoichiometry of the chemical reaction that is to be investigated. Generally, this is then followed by the measurement, at a fixed temperature, of changes in concentrations of reactants or products as a function of time. Essentially, the collection of kinetic data is an exercise in analytical chemistry with the added dimension of time. The requirement of the analytical technique — and many have been used — is that it can measure concentration. The requirements on the design of the experiment are that the analysis does not disturb the progress of the reaction and that it is done quickly so that no significant reaction occurs while it is being carried out.

Starting a reaction imposes its own constraints. If there is only one reactant, the usual procedure is to introduce the reactant into a reaction vessel which has been heated (or cooled) to the desired reaction temperature. This can also be done when there are two or more reactants, as long as they are gases, because gases can be heated (or cooled) relatively quickly. For liquid reactants, or reactants in solution, best practice is to bring the separate reactants to the reaction temperature and *then* mix them in a reaction vessel which is itself held at the desired temperature under some form of thermostatic control. The key point is that both heating (or cooling) and mixing processes take a finite time, typically of the order of seconds. To obtain meaningful kinetic data, therefore, the time for the reaction being studied to go to completion must be long compared to the time taken for the reactants to mix and reach the reaction temperature. This condition is satisfied by reactions that are referred to as being *slow*; the lower limit of time for such reactions to reach completion is about a minute or so. Slow reactions, which are our main concern here, can be studied using *conventional techniques*. Special techniques, including the use of flow systems, are required to monitor faster reactions.

The simplest conventional procedures for chemical kinetic investigations are based on chemical methods of analysis although these are now being largely superseded by more modern approaches (see below). However, the ideas underlying their application still remain instructive. The efficient way to use a chemical method of analysis is to withdraw small samples from the reaction mixture at selected times for analysis.

● Will this affect the reaction mixture in any way?

● The only effect will be to reduce the *mass* of the reaction mixture. The concentration of the reaction mixture and, hence, the rate of reaction will not be affected.

Of course, for the strategy of taking small samples from the reaction mixture to be successful, the time taken for sampling and analysis must be very short compared with the time during which significant changes in the composition of the reaction mixture occur. To avoid this particular time limitation, it is best to **quench**, or stop, the reaction immediately after the sample is taken.

Can you suggest a method of quenching the reaction?

One method would be to cool the sample rapidly to a temperature at which the reaction rate was very slow. (Often an ice bath can be used for this purpose.)

An alternative would be to use an approach in which a known amount of some reagent, which reacts rapidly with one of the reactants, is added in excess so that this reactant is completely removed from the sample taken. From a determination of the reagent that is left over it is then possible to work back and find the concentration of reactant in the original sample.

Sampling is time-consuming and it is preferable to monitor the concentration of a reactant or product *directly* as the reaction proceeds; this requires a **physical method of analysis**.

Physical methods rely on the measurement of a property of the reaction mixture that can be related to the concentration of a reactant or product species. Properties that have been widely used are pressure (in the case of gas-phase reactions), spectrophotometric absorption and electrical conductivity. Spectroscopic methods such as nuclear magnetic resonance (NMR) and infrared (IR) may also be used to analyse a mixture as the reaction progresses. From the point of view of slow reactions, the time required to make a measurement for any of these physical methods, that is the *response time*, is minimal. Physical methods of analysis are now widely used in chemical kinetic investigations.

5.2 A strategy

To establish the form of an experimental rate equation it is necessary to determine the values of *both* the partial orders of reaction and the experimental rate constant. There is no definitive set of rules for carrying out this process and, for example, a particular approach may be influenced by knowledge gained about the kinetic behaviour of similar reactions. In any approach, however, there are common steps and one strategy based on these steps is shown as a flow diagram in Figure 5.1. This, in outline, is the strategy we shall adopt here. It is assumed that the stoichiometry of the reaction has been established and that kinetic reaction profiles are to be measured at a fixed temperature.

In the sections that follow we shall take a closer look at the steps in the flow diagram. It should be emphasized, however, that the diagram is no more than an outline. Depending on circumstances, different routes can be taken to gain information about experimental rate equations and we shall mention some of these in our discussion. One key strategic point is to distinguish between reactions that involve either a single reactant or several reactants. The case of a reaction involving a single reactant provides the best starting point for discussion.

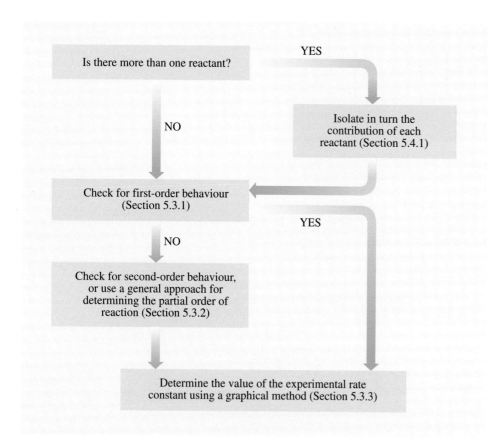

5.3 Reactions involving a single reactant

5.3.1 A preliminary half-life check

The thermal decomposition of dinitrogen pentoxide (N_2O_5) in the gas phase has time-independent stoichiometry

$$2N_2O_5(g) = 4NO_2(g) + O_2(g) \tag{5.1}$$

A kinetic reaction profile for N_2O_5 measured at 63.3 °C is shown in Figure 5.2. (If you wish to plot this kinetic reaction profile for yourself using the *Kinetics Toolkit*, the data are given in Table 5.1.)

It would be convenient if the kinetic reaction profile in Figure 5.2 could be used directly, without the need for any further processing of the data, to obtain information about the experimental rate equation for the decomposition of N_2O_5. In fact, a *preliminary check* can be carried out using a method based on the idea of **reaction half-life**, which is denoted by $t_{1/2}$. This approach was suggested many years ago by Wilhelm Ostwald who was Professor of Chemistry at Leipzig (1887–1906) and a Nobel prizewinner (1909).

The half-life for a chemical reaction involving a single reactant can be defined as: *the time it takes for the concentration of the reactant to fall to one-half of its initial value.* (It is important to note that it is *not* equivalent to one-half the time required for the reaction to go to completion.)

Table 5.1

Data for plotting the kinetic reaction profile for N_2O_5 in Figure 5.2

| time | $[N_2O_5]$ |
s	$mol\,dm^{-3}$
0	4.00×10^{-3}
50	3.25×10^{-3}
110	2.54×10^{-3}
160	2.06×10^{-3}
220	1.61×10^{-3}
330	1.02×10^{-3}
520	0.46×10^{-3}
670	0.25×10^{-3}
880	0.10×10^{-3}

What is the half-life for the decomposition of N_2O_5 at 63.3 °C?

To determine $t_{1/2}$, the time taken for the initial concentration of N_2O_5 (4.0×10^{-3} mol dm^{-3} as taken from Table 5.1) to fall to one-half of its value, that is 2.0×10^{-3} mol dm^{-3}, is required. According to Figure 5.2 this time is close to 165 s; so $t_{1/2} \approx 165$ s.

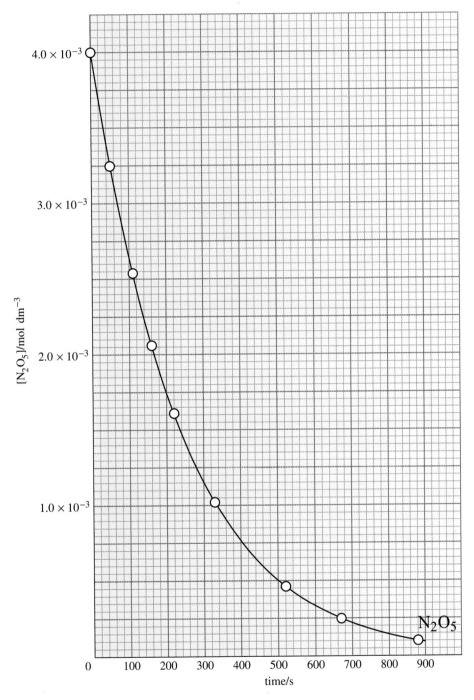

Figure 5.2
A kinetic reaction profile for N_2O_5 measured for the decomposition of this compound in the gas phase at 63.3 °C. A smooth curve is drawn through the experimental data points.

For the purposes of our preliminary check it is useful to extend the idea of reaction half-life a little further. As shown schematically in Figure 5.3, *successive half-lives* can be defined on the *same* kinetic reaction profile of a reactant A with initial concentration $[A]_0$:

- the first half-life, $t_{1/2}(1)$, corresponds to the time taken for the initial concentration to fall to $\frac{1}{2}[A]_0$,
- the second half-life, $t_{1/2}(2)$, corresponds to the time taken for the concentration to fall from $\frac{1}{2}[A]_0$ to $\frac{1}{4}[A]_0$,
- the third half-life, $t_{1/2}(3)$, corresponds to the time taken for the concentration to fall from $\frac{1}{4}[A]_0$ to $\frac{1}{8}[A]_0$.

There is no need to consider the fourth half-life, and so on.

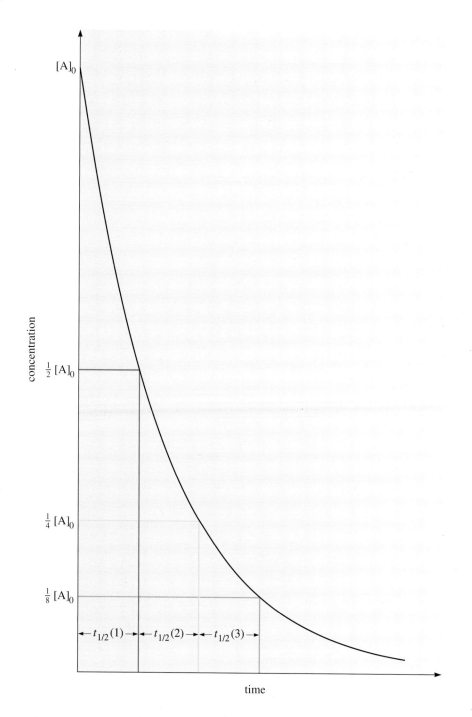

Figure 5.3
Successive half-lives for a reaction involving a single reactant A with initial concentration $[A]_0$.

⬤ What are the successive half-lives for the decomposition of N_2O_5 at 63.3 °C?

⬤ From Figure 5.2, the successive half-lives are equal to one another, each having a value close to 165 s.

The observation that successive half-lives are equal to one another is important because such behaviour is *unique* to first-order reactions. The experimental rate equation for the decomposition of N_2O_5 at 63.3 °C must therefore be

$$J = k_R[N_2O_5] \tag{5.2}$$

As indicated in the flow diagram (Figure 5.1) determination of the experimental rate constant requires a further step, although we shall not discuss this until Section 5.3.3.

> *To summarize* It is useful to check whether successive half-lives for a reaction are equal to one another, since, if this is the case, then it is safe to conclude that the reaction is first-order. In order for the check to be carried out, data must be available over at least two, preferably three, half-lives.

5.3.2 The differential method

This method has been used for many years since it was first suggested in 1884 by the Dutch chemist and first Nobel prizewinner in chemistry (1901) Jacobus Henricus van't Hoff who made substantial contributions in both physical chemistry and stereochemistry. The method is sometimes named after him. Alternatively, as here, it is described by the term 'differential' which reflects the fact that, in the language of calculus, rate equations are *differential* equations. However, this is not to imply that the method involves calculus but simply to indicate that it seeks *directly* to determine the form of experimental rate equations without changing them into another form (as we shall do later in Section 5.3.3).

As an example of a reaction involving a single reactant, we can consider the gas-phase decomposition of NO_2 at 300 °C which we have already discussed in Section 3.3, Equation 3.4

$$2NO_2(g) = 2NO(g) + O_2(g)$$

A preliminary half-life check of the kinetic reaction profile for NO_2 shows that the reaction is not first-order. (You may wish to confirm this for yourself by returning to Exercise 3.1.)

To use the differential method it is necessary to propose a **plausible rate equation** and to do this the *simplest* proposal is made.

⬤ Can you suggest what this might be?

⬤ The simplest proposal would be that

$$J = k_R[NO_2]^\alpha \tag{5.3}$$

so that the rate of reaction depends *only* on the concentration of the reactant NO_2 raised to the power of the partial order α.

It is important to recognize that the plausible rate equation is only a suggestion. Nonetheless it has the attraction of relative simplicity and, according to the discussion in Section 4.2, it has a strong likelihood of being correct since it is in a form that is found *experimentally* for many chemical reactions.

The form of the proposed rate equation indicates that analysis should focus on the relationship between the rate of reaction, J, and the concentration of the reactant, $[NO_2]$.

⬤ How is J defined for the thermal decomposition of NO_2 (Reaction 3.4)?

⬤ The stoichiometric numbers are $v_{NO_2} = -2$, $v_{NO} = +2$ and $v_{O_2} = +1$; hence

$$J = -\frac{1}{2}\frac{d[NO_2]}{dt} = +\frac{1}{2}\frac{d[NO]}{dt} = +\frac{d[O_2]}{dt}$$

The rate of reaction can therefore be determined from the kinetic reaction profiles for NO_2, NO or O_2 and you should recall that you were asked to find values of J at different times for the decomposition reaction in Exercise 3.1. These values are repeated in Table 5.2 which also includes the initial rate of reaction and two further determinations at 250 s and 750 s. You should note that this table includes the values of $d[NO_2]/dt$ from which J is calculated (although values of $d[NO]/dt$ or $d[O_2]/dt$ could equally well have been used for this purpose). In addition the table gives the values of $[NO_2]$ at the selected times; these values are simply taken from the kinetic reaction profile for NO_2.

Table 5.2 Values of $[NO_2]$, $d[NO_2]/dt$ and J at selected times for the gas-phase decomposition of NO_2 at 300 °C

time	$[NO_2]$	$d[NO_2]/dt$	J
s	mol dm^{-3}	mol dm^{-3} s^{-1}	mol dm^{-3} s^{-1}
0	4.00×10^{-3}	-16.64×10^{-6}	8.32×10^{-6}
250	1.96×10^{-3}	-3.79×10^{-6}	1.90×10^{-6}
500	1.30×10^{-3}	-1.77×10^{-6}	0.88×10^{-6}
750	0.97×10^{-3}	-0.97×10^{-6}	0.48×10^{-6}
1 000	0.78×10^{-3}	-0.61×10^{-6}	0.30×10^{-6}
1 500	0.55×10^{-3}	-0.32×10^{-6}	0.16×10^{-6}

STUDY NOTE

You may wish to use the Kinetics Toolkit to confirm the values of J at 250 s and 750 s in Table 5.2. (To determine the value of the initial rate of reaction would require more experimental data for the early stages of the reaction than that given in Table 3.2.)

Table 5.2 provides all of the information necessary to proceed with the differential method. A traditional disadvantage of this method was the difficulty of determining accurate values of J from a given set of experimental data. Computer determination makes the *practical* side of the process much easier but as discussed in Section 1.1, as well as the answer to Exercise 1.1, this does not mean that the computed values of J are without uncertainties.

There are two distinct ways of using the differential method.

A check for second-order behaviour

A preliminary half-life check has shown that Reaction 3.4 is not first-order. It seems reasonable, therefore, that the next step should be to check whether it is second-order, that is

$$J = k_R[NO_2]^2 \tag{5.4}$$

This equation can be rearranged, so that

$$\frac{J}{[NO_2]^2} = k_R \tag{5.5}$$

in other words the quantity $J/[NO_2]^2$ should be constant throughout the reaction. Values of this quantity, taken at three different times in the reaction, are given in Table 5.3. Within the uncertainties inherent in the method, the values can reasonably be taken to be constant. It can thus be concluded that the thermal decomposition of NO_2 at 300 °C *does have* an experimental rate equation of the form of Equation 5.3. The partial order with respect to NO_2 is 2 and, overall, the reaction is second-order.

You will have noticed that the ratio $J/[NO_2]^2$ in Equation 5.5 is equal to the experimental rate constant. One way of determining this rate constant, therefore, would be to calculate the average value of this ratio using all of the available data. This is an acceptable approach but, on balance, it is better to use a graphical method which we shall discuss in Section 5.3.3.

Table 5.3
A check for second-order behaviour

time	$J/[NO_2]^2$
s	$dm^3\ mol^{-1}\ s^{-1}$
250	0.49
750	0.51
1 500	0.53

A general approach

This is an approach for the determination of partial order that does not involve any checking of data and which can be used to determine partial orders of *any* value, integral or otherwise. It is well suited to computer-based analysis.

If we continue with our example of the thermal decomposition of NO_2 at 300 °C (and put to one side our knowledge that it is second-order) then, as already discussed, a plausible rate equation is

$$J = k_R[NO_2]^\alpha \tag{5.3}$$

The quantity of interest in this equation is α and it appears as an exponent. To make an exponent easier to handle mathematically it is useful to take logarithms. To take logarithms of Equation 5.3 it is necessary to take logarithms of *both* sides of the equation such that

$$\ln J = \ln(k_R[NO_2]^\alpha) \tag{5.6}$$

In this case logarithms to the base e (ln) have been taken, but it would have been equally valid to select logarithms to the base 10 (log); as long as the same thing is done to both sides of the equation, there is no reason to prefer one type of logarithm to the other. By the method outlined in the accompanying Maths Help Box, Equation 5.6 simplifies to

$$\ln J = \alpha \ln([NO_2]) + \ln k_R \tag{5.7}$$

You should notice that we have written ln(quantity), for example $\ln([NO_2])$, rather than the more cumbersome, but strictly correct, ln(quantity/units), for example $\ln([NO_2]/mol\,dm^{-3})$. We shall adopt this informal practice in what follows except when giving tables of information or labelling axes on graphs.

MATHS HELP   SIMPLIFYING EQUATION 5.6

This equation can be simplified using the rules of logarithms.

Applying the multiplication rule $\ln(rs) = \ln(r) + \ln(s)$ gives

$$\ln J = \ln k_R + \ln([NO_2]^\alpha)$$

Applying the power rule $\ln(r^s) = s\ln(r)$ gives

$$\ln J = \ln k_R + \alpha \ln([NO_2])$$

The order of terms on the right-hand side can then be changed to give

$$\ln J = \alpha \ln([NO_2]) + \ln k_R$$

which corresponds to Equation 5.7.

This is in the same form as an equation for a straight line, so that

- a plot of $\ln J$ versus $\ln([NO_2])$ should be linear if the assumed form of the rate equation is correct

- the slope of the straight line will be given by α, thereby providing a value for the partial order of reaction with respect to NO_2.

A plot of $\ln(J/\text{mol dm}^{-3}\,\text{s}^{-1})$ versus $\ln([NO_2]/\text{mol dm}^{-3})$ is shown in Figure 5.4a. It is a straight line and this is emphasized by drawing on the plot the best straight line that passes through all of the data points. It is important to recognize that the fact that it is a straight line confirms that the experimental rate equation is of the form given by Equation 5.3.

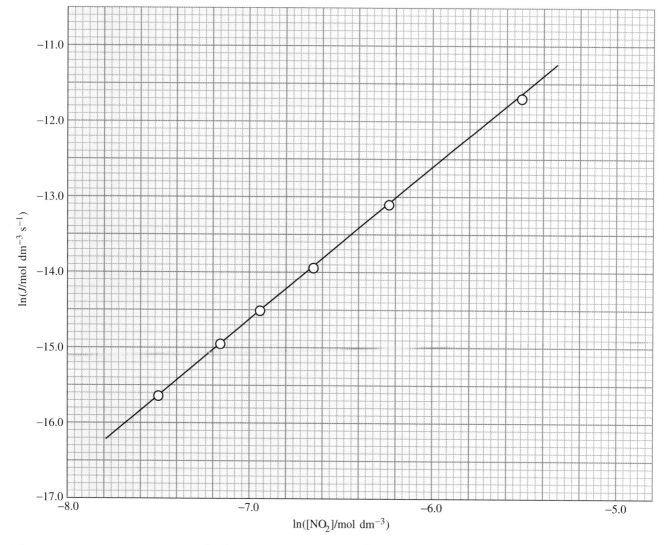

Figure 5.4(a) A plot of $\ln(J/\text{mol dm}^{-3}\,\text{s}^{-1})$ versus $\ln([NO_2]/\text{mol dm}^{-3})$ for the gas-phase thermal decomposition of NO_2 at $300°\,C$. The straight line represents the best fit to all of the data points.

QUESTION 5.1

Make a preliminary estimate of the slope of the straight line in Figure 5.4a.
Then go on to use the *Kinetics Toolkit* to determine both the slope and the
intercept.

The numerical value of the slope of the straight line can be reasonably taken to be
equal to 2, and so (as expected) the partial order with respect to NO_2 is 2.

It should be noted that since Equation 5.7 is the equation of a straight line, the
intercept on the $\ln J$ axis, which occurs when $\ln([NO_2]) = 0$, will provide a value of
$\ln k_R$, and hence a value of the experimental rate constant. However, as shown by
Figure 5.4b the data points are located in a region of the graph lying far from the
intercept and to determine this value a long *extrapolation* is required. The accuracy
of this extrapolation will depend critically on the quality of the experimental data.

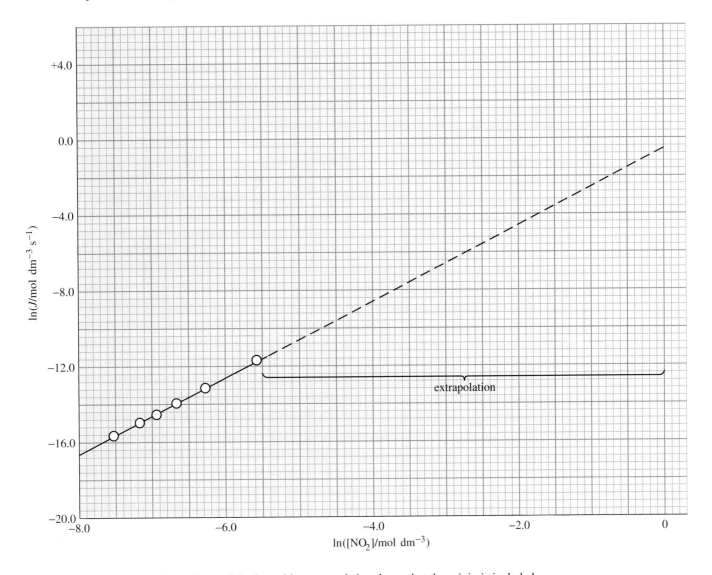

Figure 5.4(b) The same plot as Figure 5.4a, but with an extended scale, so that the origin is included.

Any small uncertainty in the slope of the line will result in a much greater uncertainty in the intercept. You will have seen that this is the case in your answer to Question 5.1. In general, the determination of an experimental rate constant from the value of an intercept calculated from a plot such as that in Figure 5.4a can introduce fairly significant uncertainties and, for this reason, it is not recommended.

To summarize We have discussed two approaches for determining partial order using the differential method. One is very specific and simply checks for second-order behaviour; in essence, it is a trial-and-error method. The other is a general graphical method which can be used to determine *any* partial order of reaction, whether it be an integer or a fraction. From the point of view of effort required, and given the availability of the *Kinetics Toolkit*, there is very little to choose between the two options. It is really a matter of preference; either carry out an initial second-order check or move straight to a more general analysis.

STUDY NOTE

By now you should have become familiar with using the Kinetics Toolkit to work through examples that are used in the text and to tackle both Questions and Exercises. Rather than keep making specific reference to using this software, we shall from now on assume that you will use it as a matter of course.

EXERCISE 5.1

The gas-phase thermal decomposition of oxygen difluoride (F_2O) in the temperature range 230 °C to 310 °C has the following time-independent stoichiometry

$$2F_2O(g) = 2F_2(g) + O_2(g)$$

Table 5.4 provides data obtained from a reaction carried out in a sealed quartz reaction vessel at 296.6 °C.

(a) Suggest a plausible rate equation for the thermal decomposition of F_2O.

(b) Plot a kinetic reaction profile for F_2O at 296.6 °C and hence demonstrate that the decomposition reaction cannot be first-order at this temperature.

(c) Using a *graphical* method, determine the partial order of reaction with respect to F_2O, and hence the overall order of reaction at 296.6 °C.

5.3.3 The integration method

A reaction involving a single reactant A

$$aA = \text{products} \tag{5.8}$$

may have an experimental rate equation of the form

$$J = -\frac{1}{a}\frac{d[A]}{dt} = k_R[A]^\alpha \tag{5.9}$$

Table 5.4

Data obtained for the gas-phase thermal decomposition of F_2O in a sealed quartz reaction vessel at 296.6 °C

time	[F_2O]
s	mol dm^{-3}
0	7.20×10^{-3}
60	5.51×10^{-3}
120	4.60×10^{-3}
180	3.83×10^{-3}
240	3.28×10^{-3}
300	2.91×10^{-3}
360	2.61×10^{-3}
420	2.36×10^{-3}
540	1.97×10^{-3}
660	1.71×10^{-3}
780	1.49×10^{-3}

Mathematically, as already mentioned in Section 5.3.2, this rate equation can be described as being a differential equation. By the application of a set of rules, it can be integrated, which means that it is converted into an alternative form. Specifically, integrating Equation 5.9 has the effect of changing the relationship between rate and concentration to one between concentration and time, and this provides a new way of analysing the kinetic data.

In the case of Equation 5.9 the result of the integration depends on the numerical value of the partial order α. However, we do not need to be concerned with the details of the integration process itself; it is the results of the process that are our main concern. The equations that result from integration are referred to as **integrated rate equations** and they are particularly important in the case of first-order ($\alpha = 1$) and second-order ($\alpha = 2$) reactions.

The integrated rate equation for a first-order reaction involving a single reactant

For a general chemical reaction of the form of Equation 5.8, the first-order experimental rate equation will be of the form

$$J = -\frac{1}{a}\frac{d[A]}{dt} = k_R[A] \tag{5.10}$$

The corresponding integrated rate equation is

$$\ln([A]_0) - \ln([A]) = ak_Rt \tag{5.11}$$

where $[A]_0$ represents the initial concentration of the single reactant A.

⬤ We have already shown that the thermal decomposition of N_2O_5

$$2N_2O_5(g) = 4NO_2(g) + O_2(g)$$

has an experimental rate equation of the form

$$J = k_R[N_2O_5]$$

What is the form of the integrated rate equation?

⬤ The integrated rate equation can be found by replacing A in Equation 5.11 by N_2O_5 and noting that $a = 2$ according to the chemical equation describing the thermal decomposition reaction. So

$$\ln([N_2O_5]_0) - \ln([N_2O_5]) = 2k_Rt \tag{5.12}$$

Equation 5.12, provides a relationship between the concentration of N_2O_5 and time. This relationship can be made clearer by rearranging the equation, so that the only term on the left-hand side is that for $\ln([N_2O_5])$

$$\ln([N_2O_5]) = -2k_Rt + \ln([N_2O_5]_0) \tag{5.13}$$

In this form, the equation shows that a plot of $\ln([N_2O_5])$ versus time (t) should be a straight line. Such a plot, based on information taken from Table 5.1, is shown in Figure 5.5; the straight line through the data points was determined as the best fit using the *Kinetics Toolkit*. It is useful to note that the observation of the good straight-line behaviour is in itself strong evidence for this decomposition reaction being first-order.

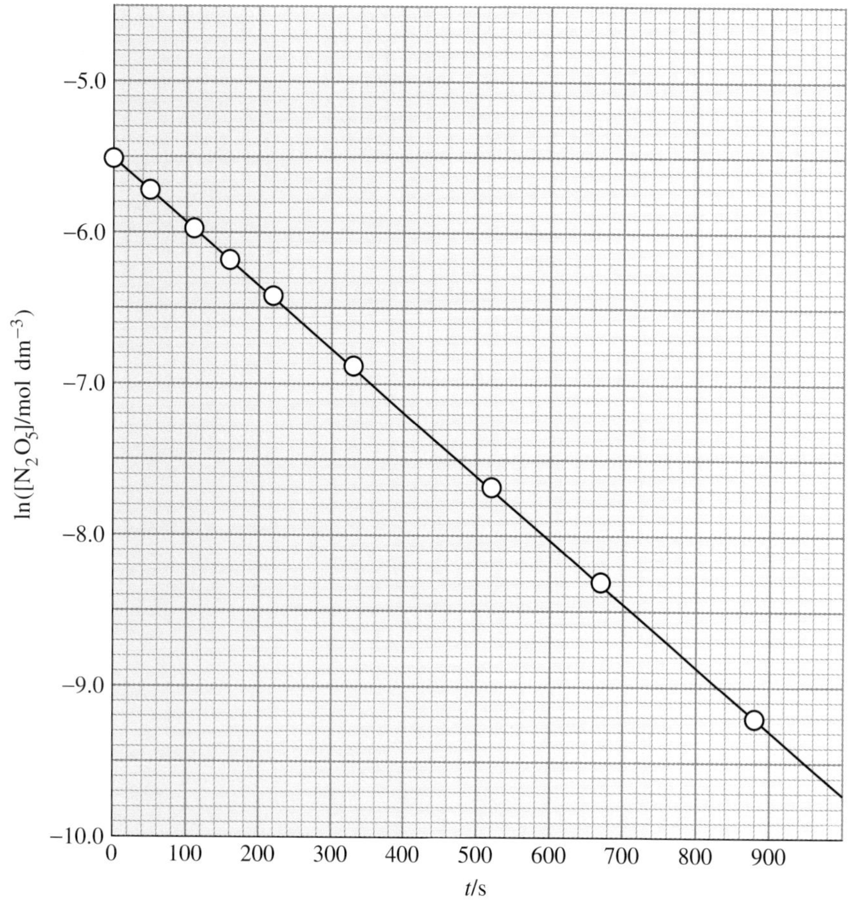

Figure 5.5
A plot of $\ln([N_2O_5]/\text{mol dm}^{-3})$ versus time (t) for the gas-phase thermal decomposition of N_2O_5 at 63.3 °C.

● If the slope of the straight line in Figure 5.5 is $-4.18 \times 10^{-3}\ \text{s}^{-1}$ (note the units), what is the value of the experimental rate constant?

● According to Equation 5.13, the slope of a plot of $\ln([N_2O_5])$ versus t will be equal to $-2k_R$. Thus

$$-2k_R = -4.18 \times 10^{-3}\ \text{s}^{-1}$$
$$k_R = 2.09 \times 10^{-3}\ \text{s}^{-1}$$

Thus, the first-order integrated rate equation provides a convenient graphical method, that uses all of the experimental data, for determining the value of the experimental rate constant. In general, depending on the quality of the experimental data that are available from a kinetic reaction profile, the uncertainty in the computed value for a given experimental rate constant can be quite small. For example, a fuller statistical analysis of the data in Figure 5.5 gives $k_R = (2.09 \pm 0.02) \times 10^{-3}\ \text{s}^{-1}$.

The integrated rate equation for a second-order reaction involving a single reactant

For a second-order experimental rate equation of the form

$$J = -\frac{1}{a}\frac{d[A]}{dt} = k_R[A]^2 \tag{5.14}$$

the corresponding integrated rate equation is

$$\frac{1}{[A]} - \frac{1}{[A]_0} = ak_R t \qquad (5.15)$$

Once again it is useful to show that this equation represents a straight line by rearranging it so that only the term $1/[A]$ is on the left-hand side

$$\frac{1}{[A]} = ak_R t + \frac{1}{[A]_0} \qquad (5.16)$$

○ What will be the slope of a plot of $1/[A]$ versus t?

○ According to Equation 5.16, a plot of $1/[A]$ versus t will be a straight line with slope = ak_R.

So, a second-order integrated rate equation also provides a convenient graphical means of determining the value of an experimental rate constant. As you can see, the method involves the use of reciprocal concentration.

MATHS HELP RECIPROCAL CONCENTRATION: LABELLING A TABLE HEADING OR AN AXIS ON A GRAPH

Suppose a reactant A has a concentration of 4.0×10^{-3} mol dm^{-3}, that is

$$[A] = 4.0 \times 10^{-3} \text{ mol dm}^{-3}$$

The reciprocal concentration is $1/[A]$, and

$$\frac{1}{[A]} = \frac{1}{4.0 \times 10^{-3} \text{ mol dm}^{-3}}$$

$$= 250 \text{ mol}^{-1} \text{ dm}^3$$

For a graph, only *pure numbers* can be plotted. If

$$\frac{1}{[A]} = 250 \text{ mol}^{-1} \text{ dm}^3$$

then, multiplying both sides by mol dm^{-3} gives

$$\frac{\text{mol dm}^{-3}}{[A]} = 250 \times \text{mol}^{-1} \text{ dm}^3 \times \text{mol dm}^{-3}$$

$$= 250$$

Thus an appropriate label for showing reciprocal concentrations (in either a table heading or on an axis of a graph) would be mol dm$^{-3}/[A]$.

QUESTION 5.2

We have shown (in Section 5.3.2) that the gas-phase thermal decomposition of NO_2 at 300 °C

$$2NO_2(g) = 2NO(g) + O_2(g)$$

has a second-order experimental rate equation

$$J = k_R[NO_2]^2$$

Use information in Table 3.2 to determine the value of the experimental rate constant k_R for this decomposition.

Determining reaction order

The most important use of integrated rate equations is in the determination of values for experimental rate constants. However, they are sometimes used in a trial-and-error procedure to determine whether a reaction is first- or second-order. The essence of the method is to see whether a good straight-line plot is obtained. Thus, for example, without any prior knowledge of the form of the experimental rate equation for the decomposition of N_2O_5, the straight line plotted in Figure 5.5, as already mentioned, would be very strong evidence that the reaction was first-order. Similar comments apply to the decomposition of NO_2; the straight line plotted in Figure Q.2 in the answer to Question 5.2 provides compelling evidence that the reaction is second-order.

There is, however, one important point to note.

> To use first- and second-order integrated rate equations in a trial-and-error procedure to determine order, it essential that the data analysed should extend to at least 50% of complete reaction, and preferably more.

The reason for this condition can be appreciated by looking at Figure 5.6. Figure 5.6a shows a *first-order* plot for the thermal decomposition of NO_2 at 300 °C. The time covered by the plot is such that the initial concentration, $[NO_2]_0 = 4.00 \times 10^{-3}$ mol dm^{-3}, falls to approximately 2.30×10^{-3} mol dm^{-3} after 180 s of reaction. A reasonable straight line can be drawn through the data points. However, as we know, the reaction is *not* first-order and this point is seen more clearly in Figure 5.6b which shows that a distinctly curved plot is obtained when data over a much longer time of reaction are considered.

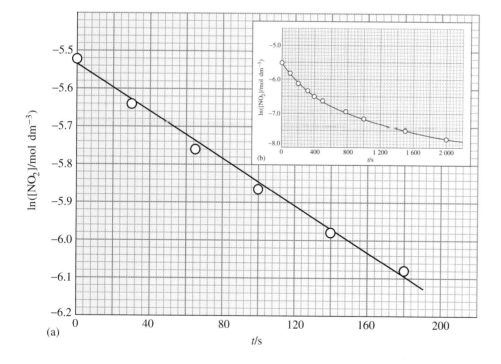

Figure 5.6 (a) A plot of $\ln([NO_2]/\text{mol dm}^{-3})$ versus time (t), for the gas-phase thermal decomposition of NO_2 at 300 °C for the first 180 s of reaction. (b, inset) The same plot on a time extended to 2 000 s.

To summarize Integrated rate equations provide a convenient graphical means, taking into account all of the experimental data, for the determination of the values of experimental rate constants. It is our preferred approach for this purpose. First- and second-order integrated rate equations can be used on a trial-and-error basis to establish order but the experimental data must extend to at least 50% of complete reaction.

Table 5.5
Data obtained for the thermal decomposition of azomethane in the gas phase at 332.4 °C

time	$[CH_3N_2CH_3]$
s	mol dm^{-3}
0	1.60×10^{-3}
240	1.12×10^{-3}
360	0.94×10^{-3}
540	0.72×10^{-3}
780	0.50×10^{-3}
1 020	0.35×10^{-3}
1 260	0.25×10^{-3}
1 620	0.15×10^{-3}
2 040	0.08×10^{-3}
2 400	0.05×10^{-3}

EXERCISE 5.2

Azomethane ($CH_3N_2CH_3$) undergoes thermal decomposition in the gas phase at 332.4 °C and the reaction can be represented by the following time-independent stoichiometry

$$CH_3N_2CH_3(g) = 2CH_3{}^\bullet(g) + N_2(g)$$

where the species $CH_3{}^\bullet$, in which the carbon atom is not fully bonded, is a methyl radical. Table 5.5 provides experimental data for the decomposition reaction.

Use the information in Table 5.5 to determine the form of the experimental rate equation, *including the value of the experimental rate constant*, for the decomposition reaction of azomethane at 332.4 °C. You should use the most convenient method of analysis for the given data.

5.4 Reactions involving several reactants

The majority of chemical reactions involve not one but several reactants and so it is important to consider how to establish the form of an experimental rate equation in these circumstances.

If we consider a general reaction between two reactants A and B

$$aA + bB = \text{products} \tag{5.17}$$

then we need to be able to establish whether the rate equation is of the form

$$J = k_R[A]^\alpha[B]^\beta \tag{5.18}$$

The problem is that the rate of reaction now depends upon the concentrations of *both* reactants and, as a consequence, it is difficult to disentangle the effect of one from the other. If there is a third reactant then the situation becomes even more complex. The solution to the problem is to arrange matters *experimentally* so that the analysis of the kinetic data can be simplified. There are two ways to achieve this. The first is quite general and is referred to as an **isolation method**. The other is more restricted in that it only applies to the initial stages of a reaction; it is referred to as the **initial rate method**.

5.4.1 The isolation method

One of the main examples we used in Section 3 was the reaction between hypochlorite ion and bromide ion in aqueous solution

$$ClO^-(aq) + Br^-(aq) = BrO^-(aq) + Cl^-(aq) \tag{3.1}$$

What would be a plausible rate equation for this reaction?

⬤ It would be of the form

$$J = k_R[ClO^-]^\alpha[Br^-]^\beta \tag{5.19}$$

The kinetic reaction profile (Figure 3.1) we discussed for this reaction was one for which the initial concentrations of ClO^- and Br^- were relatively small and similar in magnitude (3.230×10^{-3} mol dm^{-3} and 2.508×10^{-3} mol dm^{-3}, respectively). However, the reaction can be investigated over a much wider range of reactant concentrations; a specific example is shown in Figure 5.7. (The temperature is reduced to 15 °C so that the reaction is slow enough to be monitored by conventional techniques.)

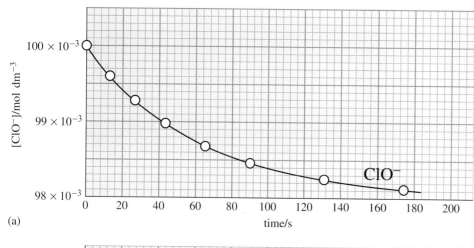

(a)

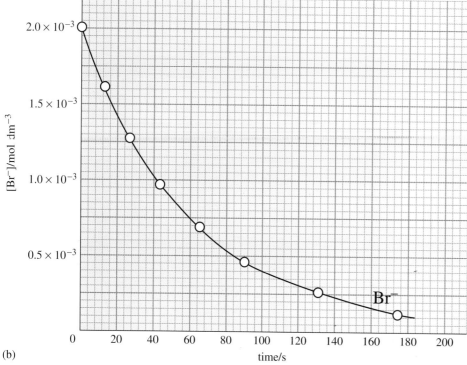

(b)

Figure 5.7   Kinetic reaction profiles for (a) ClO^- and (b) Br^- for the reaction between these ionic species in aqueous solution at 15 °C. In each plot, a smooth curve has been drawn through the experimental data points.

The initial concentration of ClO^- is $[ClO^-]_0 = 0.1$ mol dm^{-3} (although this is shown in Figure 5.7a as 100×10^{-3} mol dm^{-3} in order to facilitate comparison with the kinetic reaction profile for Br^- in Figure 5.7b). It is considerably in excess of that for Br^-: $[Br^-]_0 = 2.00 \times 10^{-3}$ mol dm^{-3}. In fact, $[ClO^-]_0/[Br^-]_0 = 50$ and so ClO^- is referred to as being in *fifty-fold excess*. The consequence of this large excess can clearly be seen in Figure 5.7; during the course of reaction Br^- is nearly all consumed, whereas most of the ClO^- remains unreacted. This being the case it is a reasonable approximation to treat the concentration of ClO^- as *remaining constant at its initial value*, $[ClO^-]_0$, throughout the course of the reaction. Under these circumstances, the proposed experimental rate equation (Equation 5.19) becomes

$$J = k_R[ClO^-]_0{}^\alpha[Br^-]^\beta \tag{5.20}$$

and the term $k_R[ClO^-]_0{}^\alpha$, which is effectively a constant, is referred to as a **pseudo-order rate constant**. It is represented by k_R', so that

$$k_R' = k_R[ClO^-]_0{}^\alpha \tag{5.21}$$

The proposed experimental rate equation therefore becomes

$$J = k_R'[Br^-]^\beta \tag{5.22}$$

The crucial point about this equation is that the rate of reaction now depends *only* on the concentration of Br^-; in other words the kinetic contribution of this reactant has been *isolated* by arranging for the concentration of ClO^- to be in large excess. The form of the expression is exactly the same as that which would be proposed for a reaction involving a single reactant. Hence, the various methods available for determining the partial order for a single-reactant process (Section 5.3) can be applied in an identical way in order to determine β.

● By referring to the kinetic reaction profile in Figure 5.7b, determine a value for β.

○ The first three successive half-lives are very close in value (approximately 42 s in each case) which is typical of a first-order reaction; thus $\beta = 1$.

Equation 5.22 is therefore referred to as a **pseudo-order rate equation**; in fact it is pseudo-first-order.

QUESTION 5.3

Table 5.6 provides information taken from the kinetic reaction profile for Br^- in Figure 5.7b. Use this information to determine a value for the pseudo-order rate constant k_R' in Equation 5.22.

To determine the partial order with respect to ClO^- (that is α in Equation 5.19) it is necessary to carry out a second experiment but this time with Br^- in large excess, that is

$$[Br^-]_0 \gg [ClO^-]_0$$

In this case, the concentration of Br^- is treated as remaining constant throughout the course of the reaction so that

$$J = k_R''[ClO^-]^\alpha \tag{5.23}$$

where $k_R'' = k_R[Br^-]_0{}^\beta$ is another pseudo-order rate constant. We have already established that $\beta = 1$. In fact, although we shall not go into detail, it turns out that $\alpha = 1$ and so the experimental rate equation for the reaction between ClO^- and Br^- in aqueous solution is

$$J = k_R[ClO^-][Br^-] \tag{5.24}$$

Table 5.6

Values of $[Br^-]$ as a function of time taken from the kinetic reaction profile for Br^- in Figure 5.7b

time	$[Br^-]$
s	mol dm^{-3}
0	2.00×10^{-3}
15	1.57×10^{-3}
28	1.27×10^{-3}
44	0.98×10^{-3}
70	0.64×10^{-3}
91	0.46×10^{-3}
130	0.25×10^{-3}

One option for determining the value of the experimental second-order rate constant, k_R, is to return to Equation 5.21. Rearranging this equation gives

$$k_R = \frac{k_R'}{[ClO^-]_0{}^\alpha} \qquad (5.25)$$

Values of k_R' (determined in Question 5.3), $[ClO^-]_0$ and α are known and so, therefore

$$k_R = \frac{1.60 \times 10^{-2} \text{ s}^{-1}}{0.1 \text{ mol dm}^{-3}}$$
$$= 0.16 \text{ dm}^3 \text{ mol}^{-1} \text{ s}^{-1}$$

> *To summarize* The isolation method provides a very valuable means of investigating the chemical kinetics of reactions that involve two, or more, reactants. In essence, it involves *isolating* in turn the contribution of each reactant by arranging (experimentally) that all of the other reactants are in *large excess*, such that their concentrations remain virtually unchanged during the course of reaction. Normally this means at least a ten-fold, but more preferably forty-fold or more, excess in concentration compared with the initial concentration of the reactant to be isolated.

It is worth noting that some types of reaction automatically have one reactant in large excess. Thus, for example, benzenediazonium chloride ($C_6H_5N_2Cl$) decomposes in water at 40 °C to liberate nitrogen gas

$$C_6H_5N_2Cl(aq) + H_2O(l) = C_6H_5OH(aq) + N_2(g) + HCl(aq) \qquad (5.26)$$

In this case water is the solvent, as well as a reactant, and it remains in large excess throughout the reaction. In this, and similar circumstances, the experimental rate equation is often written in a way that does not explicitly take the water into account. So for Reaction 5.26, the experimental rate equation would be proposed to be

$$J = k_R[C_6H_5N_2Cl]^\alpha \qquad (5.27)$$

where it is understood that k_R is a pseudo-order rate constant.

QUESTION 5.4

The reaction between the amine, $CH_2{=}CHCH_2NH_2$, and the anhydride, $(C_6H_5CH{=}CHCO)_2O$, takes place under particular conditions at a fixed temperature as follows

$$CH_2{=}CHCH_2NH_2 + (C_6H_5CH{=}CHCO)_2O$$
$$= C_6H_5CH{=}CHCOOH + C_6H_5CH{=}CHCONHCH_2CH{=}CH_2 \qquad (5.28)$$

For the purposes of this question it is convenient to represent the amine by Am and the anhydride by An. Table 5.7 provides data from an experiment carried out with the initial concentration of the amine, $[Am]_0$, equal to 1.12×10^{-4} mol dm^{-3}. Use the information in the table to determine the partial order with respect to the anhydride.

Table 5.7

Data from an experiment involving Reaction 5.28 for particular initial concentrations of the reactants

time	[An]	J
s	mol dm^{-3}	mol dm^{-3} s^{-1}
0	3.00×10^{-6}	11.23×10^{-9}
63	2.33×10^{-6}	8.73×10^{-9}

5.4.2 The initial rate method

This method, as its name implies, focuses attention on the initial stages of a reaction and it is initial rates of reaction that are measured. Experimental factors that must be taken into account when measuring these rates have already been discussed at the end of Section 3.2. If a reaction has time-independent stoichiometry then it is reasonable to assume that the kinetic information obtained from investigating the initial stages of the reaction will apply throughout its whole course. Alternatively, if a reaction forms products that decompose, or interfere in some way with the progress of a reaction, then the initial rate method may be the only viable means of obtaining useful kinetic information. The method is not included in the flow diagram in Figure 5.1 since it is essentially a stand-alone approach.

If we return to a general reaction between two reactants (Reaction 5.17)

$$a\text{A} + b\text{B} = \text{products}$$

then a plausible rate equation is (Reaction 5.18)

$$J = k_R[\text{A}]^\alpha[\text{B}]^\beta$$

and the *initial* rate of reaction, J_0, can be expressed as

$$J_0 = k_R[\text{A}]_0^\alpha[\text{B}]_0^\beta \qquad (5.29)$$

In this expression $[\text{A}]_0$ and $[\text{B}]_0$ are the initial concentrations of the reactants A and B. Their values can be pre-selected before the start of an experiment. It is thus possible to imagine carrying out a series of experiments in which, say, the initial concentration of A is fixed but the initial concentration of B is varied. For this particular series of experiments, therefore, the proposed rate equation will be

$$J_0 = k_R'[\text{B}]_0^\beta \qquad (5.30)$$

where $k_R' = k_R[\text{A}]_0^\alpha$ and is a pseudo-order rate constant. It is important to re-emphasize that it is the fact that the experimental value of $[\text{A}]_0$ is held *fixed* that allows the simplification of the proposed rate equation. In effect, this is an isolation technique because the dependence of the initial rate on just the initial concentration of B can now be investigated.

The method can be illustrated by considering results for the reaction between hypochlorite ion and bromide ion in aqueous solution (Reaction 3.1)

$$\text{ClO}^-(\text{aq}) + \text{Br}^-(\text{aq}) = \text{BrO}^-(\text{aq}) + \text{Cl}^-(\text{aq})$$

These are given in Table 5.8.

Table 5.8 Data for the initial rate method for the reaction between hypochlorite ion and bromide ion in aqueous solution (Reaction 3.1) at 25 °C

	$\dfrac{[\text{ClO}^-]_0}{\text{mol dm}^{-3}}$	$\dfrac{[\text{Br}^-]_0}{\text{mol dm}^{-3}}$	$\dfrac{J_0}{\text{mol dm}^{-3}\,\text{s}^{-1}}$
Experiment 1	3.230×10^{-3}	2.508×10^{-3}	3.19×10^{-6}
Experiment 2	6.070×10^{-3}	2.507×10^{-3}	5.98×10^{-6}
Experiment 3	9.253×10^{-3}	2.509×10^{-3}	9.14×10^{-6}

You may have noticed in Table 5.8 that the initial concentrations of ClO$^-$ and Br$^-$ in *Experiment 1* are the same as those used in plotting the kinetic reaction profile in Figure 3.1 (Table 3.1, Section 3.1). In all three experiments, at least within experimental error, the initial concentration of Br$^-$ is fixed, *but it is not in excess*. This is an important feature of the initial rate method: it is not necessary to have reactants in excess.

⬤ What will be the form of a plausible rate equation for the series of experiments in Table 5.8?

⬤ The plausible rate equation for the initial reaction will be

$$J_0 = k_R[ClO^-]_0{}^\alpha[Br^-]_0{}^\beta \qquad (5.31)$$

However, since $[Br^-]_0$ is fixed in all three experiments, then

$$J_0 = k_R'[ClO^-]_0{}^\alpha \qquad (5.32)$$

where $k_R' = k_R[Br^-]_0{}^\beta$ and is a pseudo-order rate constant.

⬤ Are the data in Table 5.8 consistent with the result that the partial order with respect to ClO^- in Reaction 3.1 is equal to 1?

⬤ If $\alpha = 1$, then in all three experiments the quantity $J_0/[ClO^-]_0$ should be, within experimental uncertainty, equal to a constant value (that is, k_R'). Thus in the case of *Experiment 1*

$$\frac{J_0}{[ClO^-]_0} = \frac{3.19 \times 10^{-6} \text{ mol dm}^{-3} \text{ s}^{-1}}{3.230 \times 10^{-3} \text{ mol dm}^{-3}}$$

$$= 9.88 \times 10^{-4} \text{ s}^{-1}$$

The corresponding values for *Experiments 2* and *3* are 9.85×10^{-4} s^{-1} and 9.88×10^{-4} s^{-1}, respectively. Thus in all three experiments the value of $J_0/[ClO^-]_0$ is essentially constant.

If the partial order with respect to ClO^- had not been equal to 1 then it would have been necessary to proceed by a trial-and-error method in order to find the value of α that made $J_0/[ClO^-]_0{}^\alpha$ equal to the same constant value for all three experiments.

> *To summarize*   The initial rate method is essentially an isolation technique but it does not require that any reactants have to be in large excess. In general, for a reaction involving two or more reactants, one of these is isolated by arranging that the initial concentrations of the others are held at fixed values during a series of experiments. The main application of the method is for the determination of partial order. Values of pseudo-order rate constants can be determined but with an accuracy that, in turn, depends on how accurately initial rates of reaction can be measured.

QUESTION 5.5

Further information for the reaction between an amine (Am) and an anhydride (An), described in Question 5.4, is given in Table 5.9. Use this information to determine the partial order with respect to the amine.

Table 5.9   Further information for the reaction between an amine and an anhydride

	$\dfrac{[Am]_0}{\text{mol dm}^{-3}}$	$\dfrac{[An]_0}{\text{mol dm}^{-3}}$	$\dfrac{J_0}{\text{mol dm}^{-3} \text{ s}^{-1}}$
Experiment 1	1.12×10^{-4}	3.00×10^{-6}	11.23×10^{-9}
Experiment 2	3.83×10^{-4}	3.00×10^{-6}	38.62×10^{-9}

5.5 Summary of Section 5

1   An analytical technique that can measure concentration is required to investigate the chemical kinetics of a reaction. Such a technique must not interfere with the progress of the reaction and no significant reaction should occur while the analysis is being carried out. If samples are drawn from a reaction mixture during reaction, then it is preferable to quench them so that they can be analysed at a later time.

2   Slow reactions are those for which the lower limit of time to reach completion is of the order of a minute; such reactions can be investigated by conventional experimental techniques.

3   Physical methods of analysis, which are now widely used, allow the concentration of a reactant or product to be monitored directly and so avoid the need for sampling.

4   To establish the form of an experimental rate equation it is necessary to determine the values of both the partial orders of reaction and the experimental rate constant. There is no one set of rules for carrying out this process but a suggested strategy is given in the flow diagram in Figure 5.1.

5   The half-life for a chemical reaction involving a single reactant is defined as the time taken for the concentration of this reactant to fall to one-half of its initial value. Successive half-lives for a reaction that is first-order (or pseudo-first-order due to the use of the isolation method) are always equal to one another. The measurement of successive half-lives, therefore, provides a useful preliminary check for first-order behaviour.

6   For a reaction involving a single reactant A

$$a\text{A} = \text{products}$$

using the differential method involves proposing a plausible rate equation

$$J = k_\text{R}[\text{A}]^\alpha$$

and then either

(i)   checking for second-order behaviour ($\alpha = 2$) by determining whether $J/[\text{A}]^2$ has a constant value; or

(ii)   using a direct graphical approach and determining the value of α from a plot of $\ln J$ versus $\ln([A])$.

In either case the value of the experimental rate constant is best determined using a graphical method based on the appropriate integrated rate equation.

7   For a reaction involving a single reactant the first-order and second-order integrated rate equations are summarized in Table 5.10 (overleaf). Each equation is written in a form such that a straight-line graph can be plotted. In each case the slope of the straight line is directly related to the value of the experimental rate constant. Integrated rate equations can be used to determine the order of reaction but for this purpose the data to be analysed must extend to 50% of complete reaction, and preferably more.

Table 5.10 First-order and second-order integrated rate equations for a reaction of the form $a\text{A}$ = products

Reaction order	Rate equation	Integrated rate equation
first	$-\dfrac{1}{a}\dfrac{d[\text{A}]}{dt} = k_R[\text{A}]$	$\ln([\text{A}]) = -ak_R t + \ln([\text{A}]_0)$ graphical plotting: $\ln([\text{A}])$ versus t; slope $= -ak_R$
second	$-\dfrac{1}{a}\dfrac{d[\text{A}]}{dt} = k_R[\text{A}]^2$	$\dfrac{1}{[\text{A}]} = ak_R t + \dfrac{1}{[\text{A}]_0}$ graphical plotting: $\dfrac{1}{[\text{A}]}$ versus t; slope $= ak_R$

8   For reactions involving several reactants it is convenient to arrange matters experimentally so that the analysis of the kinetic data can be simplified. One general approach is to use an isolation method such that all reactants, except the one of interest, are in large excess, that is at least ten-fold but preferably forty-fold or more. Alternatively, an initial rate method can be used in which one reactant is isolated by arranging that the initial concentrations of all of the other reactants are held at fixed values, but not necessarily excess values, in a series of experiments.

THE EFFECT OF TEMPERATURE ON THE RATE OF A CHEMICAL REACTION

6

6.1 The Arrhenius equation

As we have seen, in order to determine the form of the experimental rate equation for a chemical reaction it is necessary to carry out experiments at a fixed temperature. This is to avoid any complications due to the rate of reaction changing as a function of temperature. In general, it is the rate constant, k_R, for a chemical reaction that is temperature-dependent and this is illustrated in Figure 6.1 for the reaction between iodomethane (CH_3I) and ethoxide ion ($C_2H_5O^-$) in a solution of ethanol

$$CH_3I + C_2H_5O^- = C_2H_5OCH_3 + I^- \tag{6.1}$$

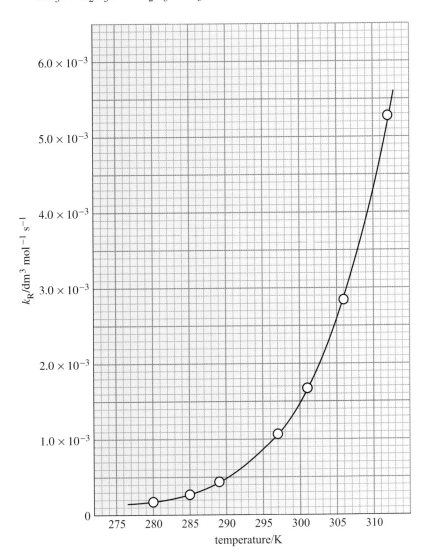

Figure 6.1
The variation of the rate constant for Reaction 6.1 with temperature. A smooth curve has been drawn through the experimental data points.

Reaction 6.1 is second-order overall, that is the experimental rate equation is of the form

$$J = k_R[CH_3I][C_2H_5O^-] \qquad (6.2)$$

and so, as shown in Figure 6.1 the units of k_R are $dm^3\,mol^{-1}\,s^{-1}$. It is very important to note in the figure that temperatures are expressed on the Kelvin scale, rather than the Celsius scale. It is often the case that the Celsius scale is used to report the temperature at which an experimental kinetic study is carried out. However, for any subsequent calculations in which temperature is involved it is *essential* to convert to kelvin (K); it is tempting to use the word *degree* in this unit name, but this would be incorrect, as would be the inclusion of the degree symbol ($°$).

For a period of over 60 years, spanning 1850 to 1910, in spite of considerable experimental effort, there was much uncertainty and, in some cases, controversy as to how to describe the temperature-dependent behaviour of the kinetics of a chemical reaction. During this period a number of empirical relationships linking the experimental rate constant to temperature were proposed and championed by different groups of workers. The situation is aptly described in a famous textbook published in 1904.

'The influence of temperature on chemical reactions is so very marked that this has been universally recognized as one of the most important factors in the study of chemical changes. Although many interesting facts have been brought to light by a happy combination of theory and experiment, this subject still forms, as Ostwald has said, "one of the darkest chapters in chemical mechanics." The subject, too, is vested with a certain amount of technical interest, since the manufacturer must know the best temperature to keep unstable solutions, such as the "azo-colours" of the dye-house, in order to have a minimum loss by decomposition.'

Chemical Statics and Dynamics by J. W. Mellor (1904)

However, from about 1910 onwards the equation that gained general acceptance was the one used by Svante Arrhenius in a paper entitled, '*On the reaction velocity of the inversion of cane sugar by acids*', published in 1889. The equation, which is now referred to as the **Arrhenius equation**, takes the form

$$k_R = A\exp\left(-\frac{E_a}{RT}\right) \qquad (6.3)$$

so that an *exponential dependence* is involved in relating the experimental rate constant to temperature (T). (In this equation we have chosen to represent the exponential of the term ($-E_a/RT$) by using 'exp' rather than 'e'; both representations are commonly used.)

In the Arrhenius equation the parameter E_a is known as the **Arrhenius activation energy** but usually is just referred to as the *activation energy*. The parameter A is the **Arrhenius A-factor** but, again, this is usually shortened to just A-*factor*; it is worth noting that the terms *pre-exponential factor* or *frequency factor* are sometimes used. Together, the two parameters A and E_a are known as the **Arrhenius parameters**. The quantity R is the gas constant; $R = 8.314\,41\,J\,K^{-1}\,mol^{-1}$ (although we shall take the value to just three decimal places in our calculations).

As its name implies the activation energy has units related to energy and these are normally expressed as $kJ\,mol^{-1}$. The units of the quantity RT are ($J\,K^{-1}\,mol^{-1}$) $\times$ (K); that is, $J\,mol^{-1}$.

⬤ Only a pure number (that is, a dimensionless quantity) can be used to calculate an exponential so this implies that the quantity E_a/RT has no units. Is this the case?

⬤ If the units of E_a are kJ mol^{-1} then these can be equally well be expressed in J mol^{-1} by simply multiplying the value in kJ mol^{-1} by 1 000. The units of E_a/RT are then (J mol^{-1})/(J mol^{-1}). The units cancel and E_a/RT is therefore dimensionless.

⬤ What are the units of the A-factor?

⬤ Since exp($-E_a/RT$) will be dimensionless then according to the Arrhenius equation the units of the A-factor must be the *same* as the units for the rate constant. For instance, they will be s^{-1} for a first-order reaction and dm^3 mol^{-1} s^{-1} for a second-order reaction.

An important reason for the widespread acceptance of the Arrhenius equation was that the parameters A and E_a could be given a physical meaning; as one author puts it, other suggested equations were 'theoretically sterile'. We shall look at this in a little more detail in Section 7. A very important *practical* aspect of the Arrhenius equation is that it accounts very well for the temperature-dependent behaviour of a large number of chemical reactions. In this context the Arrhenius parameters can be treated simply as experimental quantities. Even so, they provide very useful information about a given reaction, particularly in terms of the magnitude of its activation energy.

Svante Arrhenius (1859–1927)

Svante Arrhenius was born near Uppsala in Sweden. He became Professor of Physics at Stockholm in 1895 and a director of the Nobel Institute of Physical Chemistry (Stockholm) in 1905.

The equation named after him is one of the best known in chemical kinetics. However, apart from the 1889 paper, referred to earlier, he published little else in this area. His award of the third Nobel Prize for Chemistry in 1903 was in recognition of the extraordinary services he had rendered to the advancement of chemistry by his electrolytic theory of dissociation. He developed very wide research interests encompassing immunological chemistry, cosmology, the causes of the ice ages, and the origin of life.

Figure 6.2
Svante Arrhenius (1859–1927).

6.2 Determining the Arrhenius parameters

In order to determine the Arrhenius parameters for a reaction it is necessary to determine values of the experimental rate constant as a function of temperature. This set of data is then fitted to the Arrhenius equation using a graphical procedure. For many reactions, particularly in solution involving organic solvents, Arrhenius studies are restricted to differences between the melting and boiling temperatures of the solvent and this limits the number of data points that can be collected.

If logarithms to the base e (ln) are taken of Equation 6.3, then

$$\ln k_R = \ln A - \frac{E_a}{RT}$$ (6.4)

It is also common to take logarithms to the base 10 (log) and in this case

$$\log k_R = \log A - \frac{E_a}{2.303\,RT}$$ (6.5)

MATHS HELP   TAKING LOGARITHMS OF THE ARRHENIUS EQUATION

The Arrhenius equation is

$$k_R = A\,\exp\left(-\frac{E_a}{RT}\right)$$

To take logarithms of this equation it is necessary to take logarithms of both sides of the equation. It is convenient to first consider taking logarithms to the base e

$$\ln k_R = \ln\left\{A\,\exp\left(-\frac{E_a}{RT}\right)\right\}$$

If we use the multiplication rule of logarithms, then

$$\ln k_R = \ln A + \ln\left\{\exp\left(-\frac{E_a}{RT}\right)\right\}$$

An important property of an exponential is that $\ln\{\exp(x)\} = x$, where x is a number. It thus follows that

$$\ln\left\{\exp\left(-\frac{E_a}{RT}\right)\right\} = -\frac{E_a}{RT}$$

and so finally

$$\ln k_R = \ln A - \frac{E_a}{RT}$$

Logarithms to the base e are related to those to the base 10 by the following equation

$$\ln x = 2.303 \log x$$

Using logarithms to the base 10, therefore,

$$2.303 \log k_R = 2.303 \log A - \frac{E_a}{RT}$$

and if each term in this equation is divided by 2.303, then

$$\log k_R = \log A - \frac{E_a}{2.303\,RT}$$

It should be remembered that for the logarithmic terms that appear in the equations in this box, we should strictly write ln(quantity/units) or log(quantity/units).

If the order of the terms on the right-hand side of Equation 6.4 is changed, then

$$\ln k_R = -\frac{E_a}{RT} + \ln A \qquad (6.6)$$

This equation is in the form of a straight line so that a plot of $\ln k_R$ versus $1/T$ should be linear. This assumes that both A and E_a are constants, independent of temperature. This is a reasonable assumption for most reactions when studied over a limited range of temperature, say of the order of 100 K. A graph of $\ln k_R$ versus $1/T$ is referred to as an **Arrhenius plot**, and it involves *reciprocal temperature* or *inverse temperature*.

MATHS HELP   RECIPROCAL TEMPERATURE OR INVERSE TEMPERATURE

Suppose a reaction is studied at a temperature such that $T = 297$ K.

The reciprocal or inverse temperature will be

$$\frac{1}{T} = \frac{1}{297\,\text{K}}$$

$$= 3.367 \times 10^{-3}\,\text{K}^{-1}$$

To plot a graph, only pure numbers can be used. If

$$\frac{1}{T} = 3.367 \times 10^{-3}\,\text{K}^{-1}$$

then, multiplying both sides by K, gives

$$\frac{\text{K}}{T} = 3.367 \times 10^{-3}$$

Thus a label for a column heading in a table could be K/T. (In this case, the numbers that appear in the corresponding column of data are in a form suitable for *direct* entry into the graph-plotting software in the *Kinetics Toolkit*.) However, it is more conventional to include the power of ten in the label. If we return to the previous equation and multiply both sides by 10^3, then

$$10^3 \times \left(\frac{\text{K}}{T}\right) = 10^3 \times (3.367 \times 10^{-3})$$

or $\left(\dfrac{10^3\,\text{K}}{T}\right) = 3.367$

We shall use the label 10^3 K/T for the horizontal axes of Arrhenius plots.

Figure 6.3 shows an Arrhenius plot for the reaction between iodomethane and ethoxide ion in a solution of ethanol (Equation 6.1) based on the data used to plot Figure 6.1. Note, as explained in the accompanying Maths Help Box, that the horizontal axis for the Arrhenius plot is labelled 10^3 K/T. This means that any point on this axis represents the numerical value of the quantity (10^3 K/T).

Table 6.1 summarizes the data for the plots in both Figures 6.1 and 6.3. Columns of data for K/T (for direct input into the graph-plotting software of the *Kinetics Toolkit*) and 10^3 K/T (for the Arrhenius plot) are both given.

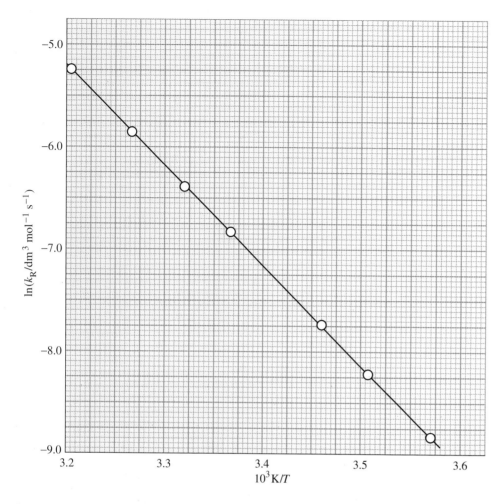

Figure 6.3
An Arrhenius plot for the reaction between iodomethane and ethoxide ion in a solution of ethanol. The straight line represents the best fit to all of the data points.

Table 6.1   Data for the Arrhenius plot in Figure 6.3

$\dfrac{T}{K}$	$\dfrac{k_R}{dm^3\ mol^{-1}\ s^{-1}}$	$\dfrac{K}{T}$	$\dfrac{10^3\ K}{T}$	$\ln\left(\dfrac{k_R}{dm^3\ mol^{-1}\ s^{-1}}\right)$
280	0.145×10^{-3}	3.571×10^{-3}	3.571	-8.839
285	0.268×10^{-3}	3.509×10^{-3}	3.509	-8.225
289	0.432×10^{-3}	3.460×10^{-3}	3.460	-7.747
297	1.078×10^{-3}	3.367×10^{-3}	3.367	-6.833
301	1.672×10^{-3}	3.322×10^{-3}	3.322	-6.394
306	2.849×10^{-3}	3.268×10^{-3}	3.268	-5.861
312	5.280×10^{-3}	3.205×10^{-3}	3.205	-5.244

The Arrhenius plot in Figure 6.3 is a good straight line.

If the coordinates for two points on this line are
$(1/T = 3.25 \times 10^{-3}\ K^{-1}, \ln(k_R/dm^3\ mol^{-1}\ s^{-1}) = -5.70)$ and
$(1/T = 3.50 \times 10^{-3}\ K^{-1}, \ln(k_R/dm^3\ mol^{-1}\ s^{-1}) = -8.15)$, what is the value of the slope?

● The slope is calculated as follows

$$\text{slope} = \frac{-8.15 - (-5.70)}{3.50 \times 10^{-3}\ \text{K}^{-1} - 3.25 \times 10^{-3}\ \text{K}^{-1}}$$

$$= \frac{-2.45}{0.25 \times 10^{-3}\ \text{K}^{-1}}$$

$$= -9.80 \times 10^{3}\ \text{K}$$

● What is the activation energy?

● According to Equation 6.6

$$\text{slope} = -\frac{E_a}{R}$$

so it follows that

$$E_a = -\text{slope} \times R$$

$$= -(-9.80 \times 10^{3}\ \text{K}) \times (8.314\ \text{J K}^{-1}\ \text{mol}^{-1})$$

$$= 81.5 \times 10^{3}\ \text{J mol}^{-1}$$

$$= 81.5\ \text{kJ mol}^{-1}$$

QUESTION 6.1

If it had been decided to plot $\log(k_R/\text{dm}^3\ \text{mol}^{-1}\ \text{s}^{-1})$ versus reciprocal temperature in order to obtain an Arrhenius plot for Reaction 6.1, what would have been the value of the slope of the straight line in this case?

If the *Kinetics Toolkit* is used in the analysis of the data in Table 6.1, then fitting to a straight line produces very good results. The computed value of the slope is $(-9.814 \pm 0.002) \times 10^{3}\ \text{K}$, so that the activation energy is calculated to be $(81.59 \pm 0.02)\ \text{kJ mol}^{-1}$.

The plausible range, determined by the fit, within which the value of the intercept can lie is also relatively small.

● According to Equation 6.6, how is the intercept related to the A-factor?

● The straight line will intersect the $\ln(k_R)$ axis when $1/T$ is zero. The intercept, therefore, is equal to $\ln A$.

The computed value of the intercept is 26.21. Since the A-factor has the same units as k_R, then

$$\ln\left(\frac{A}{\text{dm}^3\ \text{mol}^{-1}\ \text{s}^{-1}}\right) = 26.21 \tag{6.7}$$

To determine the value of $A/\text{dm}^3\ \text{mol}^{-1}\ \text{s}^{-1}$ it is necessary to take the inverse of the natural logarithm. This is equivalent to finding the exponential of the quantity on the right-hand side of Equation 6.7

$$\frac{A}{\text{dm}^3\ \text{mol}^{-1}\ \text{s}^{-1}} = \exp(26.21) \tag{6.8}$$

so that

$$\frac{A}{\text{dm}^3\ \text{mol}^{-1}\ \text{s}^{-1}} = 2.42 \times 10^{11} \tag{6.9}$$

Thus, the value of the A-factor is calculated to be $2.42 \times 10^{11}\,\text{dm}^3\,\text{mol}^{-1}\,\text{s}^{-1}$. The plausible range for this value, as estimated from the statistical information provided by the *Kinetics Toolkit*, is from $2.40 \times 10^{11}\,\text{dm}^3\,\text{mol}^{-1}\,\text{s}^{-1}$ to $2.44 \times 10^{11}\,\text{dm}^3\,\text{mol}^{-1}\,\text{s}^{-1}$. It should be noted, as shown by the extended Arrhenius plot for Reaction 6.1 in Figure 6.4, that the data points are located in a region of the graph lying far from the intercept. This behaviour is also typical of the Arrhenius plots in many other investigations. As a consequence of the relatively long extrapolations involved, any uncertainties in the slopes of such plots are translated into larger uncertainties in the determination of the intercepts and, hence, the A-factors. As a general rule, care should be taken in using the intercept of an Arrhenius plot to determine an A-factor.

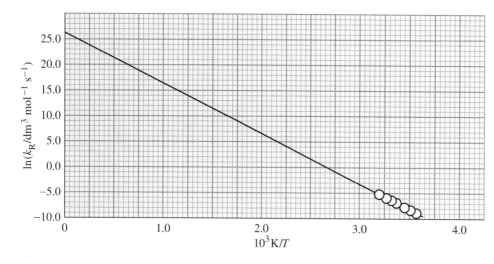

Figure 6.4
An Arrhenius plot for Reaction 6.1 on an extended scale.

QUESTION 6.2

Suppose that you were simply given the Arrhenius plot in Figure 6.3 and provided with the additional information that it was for a reaction with an activation energy of $81.59\,\text{kJ}\,\text{mol}^{-1}$. One way to determine the A-factor would be to select any point lying on the line near the middle of the data range (it need not be one of the experimental points) and determine its coordinates. This information can then be used to determine the A-factor. Try this for yourself. (A convenient point on the $10^3\,\text{K}/T$ axis is that with a numerical value equal to 3.35.)

It is useful to summarize at this stage. The Arrhenius parameters for the reaction between iodomethane and ethoxide ion in a solution of ethanol have been found to be $E_a = 81.59\,\text{kJ}\,\text{mol}^{-1}$ and $A = 2.42 \times 10^{11}\,\text{dm}^3\,\text{mol}^{-1}\,\text{s}^{-1}$. This statement is sufficient to provide a good deal of information about the influence of a change of temperature on the reaction. In particular, using this information, the value of the rate constant can be calculated at *any* temperature within the range in which the reaction was investigated. It is also, of course, possible to calculate values of the rate constant at temperatures *outside* the range of those investigated experimentally; although these temperatures should not be such that the solution would either freeze or boil. For example, the rate constant at 330 K would be calculated as follows

$$k_R = 2.42 \times 10^{11}\,\text{dm}^3\,\text{mol}^{-1}\,\text{s}^{-1} \exp\left(-\frac{81\,590\,\text{J}\,\text{mol}^{-1}}{8.314\,\text{J}\,\text{K}^{-1}\,\text{mol}^{-1} \times 330\,\text{K}}\right)$$

$$= 2.42 \times 10^{11}\,\text{dm}^3\,\text{mol}^{-1}\,\text{s}^{-1} \exp(-29.74)$$

$$= 2.94 \times 10^{-2}\,\text{dm}^3\,\text{mol}^{-1}\,\text{s}^{-1}$$

6.3 The magnitude of the activation energy

The rate of reaction *at a given temperature* for a chemical reaction with an experimental rate equation of the form given by Equation 5.18

$$J = k_R[A]^\alpha[B]^\beta$$

and an Arrhenius dependence of the rate constant on temperature given by Equation 6.3

$$k_R = A \exp\left(-\frac{E_a}{RT}\right)$$

will depend upon three key factors. These are

- the value of the *A*-factor
- the magnitude of the activation energy
- the concentrations of the reactants as determined by the initial conditions used in the experiment.

More often than not it is the magnitude of the activation energy that is the most influential of these. This can easily be appreciated by considering the data in Table 6.2. This shows how the term $\exp(-E_a/RT)$ changes as the activation energy is varied between typically low and high values at temperatures of 300 K and 600 K.

Table 6.2   Values of $\exp(-E_a/RT)$ calculated at different values of E_a at 300 K and 600 K

$\dfrac{E_a}{\text{kJ mol}^{-1}}$	$\exp\left(-\dfrac{E_a}{RT}\right)$	
	$T = 300$ K	$T = 600$ K
11.5	10^{-2}	10^{-1}
51.7	10^{-9}	3.2×10^{-5}
103.4	10^{-18}	10^{-9}

At 300 K, which is close to room temperature, a change in the magnitude of E_a by just under a factor of 10 leads to an enormous variation covering some 16 orders of magnitude. As a consequence, it is generally valid to compare the rates of *different* reactions in this temperature region solely on the basis of the magnitudes of their activation energies; any effects due to differences in *A*-factors or concentrations will be swamped by the variation in the $\exp(-E_a/RT)$ term. *As a rough generalization rapid reactions will be characterized by lower activation energies.* This generalization may still hold at higher temperatures but, as indicated in Table 6.2, the variation in the term $\exp(-E_a/RT)$ is reduced and so differences in A-factors or concentrations may become significant.

For a *particular* reaction, the magnitude of the activation energy will determine the degree to which the rate of reaction will increase when the temperature is increased under *similar* concentration conditions. If the concentration conditions are kept the same then it is only necessary to consider changes in the rate constant. If we consider two different temperatures $T(1)$ and $T(2)$, where $T(2) > T(1)$, and assume as before that A and E_a are constants independent of temperature, then using Equation 6.4 we can write

$$\ln k_R(1) = \ln A - \frac{E_a}{RT(1)} \tag{6.10}$$

and

$$\ln k_R(2) = \ln A - \frac{E_a}{RT(2)} \tag{6.11}$$

where $k_R(1)$ and $k_R(2)$ are the rate constants at temperatures $T(1)$ and $T(2)$, respectively. These equations can be combined (see the accompanying Maths Help Box)

$$\ln\left(\frac{k_R(2)}{k_R(1)}\right) = \frac{E_a}{R}\left(\frac{1}{T(1)} - \frac{1}{T(2)}\right) \tag{6.12}$$

This equation provides a means of calculating the relative change in rate constant for a reaction for a given temperature change; that is, from $T(1)$ to $T(2)$, on the basis of just a knowledge of the activation energy.

MATHS HELP   DERIVING EQUATION 6.12

This is derived by subtracting Equation 6.10 from Equation 6.11

$$\ln k_R(2) - \ln k_R(1) = \left(\ln A - \frac{E_a}{RT(2)}\right) - \left(\ln A - \frac{E_a}{RT(1)}\right)$$

On the right-hand side the two terms in $\ln A$ cancel

$$\ln k_R(2) - \ln k_R(1) = -\frac{E_a}{RT(2)} - \left(-\frac{E_a}{RT(1)}\right)$$

It is best to deal separately with each side of this equation as follows

Left-hand side

The division rule for logarithms means that

$$\ln k_R(2) - \ln k_R(1) = \ln\frac{k_R(2)}{k_R(1)}$$

Right-hand side

This can be rearranged as follows

$$-\frac{E_a}{RT(2)} - \left(-\frac{E_a}{RT(1)}\right) = -\frac{E_a}{RT(2)} + \frac{E_a}{RT(1)}$$

$$= \frac{E_a}{RT(1)} - \frac{E_a}{RT(2)}$$

$$= \frac{E_a}{R}\left(\frac{1}{T(1)} - \frac{1}{T(2)}\right)$$

Hence, overall

$$\ln\left(\frac{k_R(2)}{k_R(1)}\right) = \frac{E_a}{R}\left(\frac{1}{T(1)} - \frac{1}{T(2)}\right)$$

QUESTION 6.3

The gas-phase reaction

$$H_2C=CH-CH=CH_2 \text{ (g)}$$

which involves the ring-opening of cyclobutene to give butadiene has an activation energy of 137 kJ mol^{-1} and a rate that can be measured by conventional means at 420 K. If the temperature is increased by 10 K to 430 K, and the initial concentration of cyclobutene is kept the same, by what factor will the rate of reaction increase?

The answer to Question 6.3 illustrates a general rule-of-thumb. It can often be assumed that the rate of a reaction will increase by a factor of between 2 and 3 for each 10 K rise in temperature. In particular, for reactions taking place in aqueous solution where the temperature is limited to the range 0 °C to 100 °C, this approximation applies when the activation energy lies between about 40 kJ mol^{-1} and 130 kJ mol^{-1}. Activation energies of this order of magnitude are quite common in practice. For example, Figure 6.5 shows a histogram of the distribution of the activation energies of 147 drug decomposition reactions, mostly hydrolyses but also some oxidations and other reactions. The mean value of the activation energy is 88 kJ mol^{-1} and of all of the values, 88% fall in the range 40 kJ mol^{-1} to 130 kJ mol^{-1}.

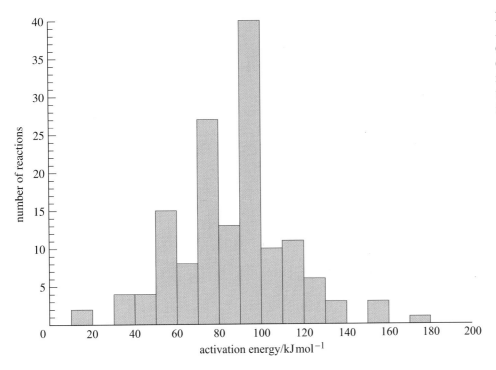

Figure 6.5
A histogram showing the distribution of the activation energies for 147 drug decomposition reactions. The activation energies on the horizontal axis are divided into increments of 10 kJ mol^{-1}.

The ability to predict changes in the rate of a reaction based on a knowledge of the magnitude of the activation energy is useful in **accelerated stability studies** for drugs in the pharmaceutical industry. Determination of the maximum safe storage period for a drug requires information on how rapidly the reactions which cause the compound to deteriorate will occur at room temperature. If the drug is to be useful, of course, such reactions must of necessity be slow but this means that kinetic measurements at around 20 °C would be very prolonged, if not totally impractical.

Very commonly therefore, values of the rate constant k_R are obtained at several temperatures in the range 40 °C to 90 °C and extrapolation is then made to room temperature. An example is provided in Question 6.4.

QUESTION 6.4

For the degradation of the antibiotic clindamycin, which is held at pH 4.0 in an aqueous solution, kinetic measurements show that at 343 K the reaction is first-order with a rate constant equal to 2.49×10^{-7} s^{-1}. Over the temperature range 320 K to 360 K, the activation energy was found to be 123.3 kJ mol^{-1}. Calculate the rate constant at 295 K. Since the degradation reaction is first-order, it turns out that the time taken for 1% decomposition turns out to be close to $0.01/k_R$ and this time is independent of the initial concentration of the antibiotic. Hence, is it possible to conclude that if the compound were to be stored at 295 K, it would remain safe for use at the end of an economically acceptable shelf-life?

To close this section it is worthwhile re-emphasizing that the Arrhenius equation is exceptional in that it accounts so well for the temperature behaviour of the vast majority of reactions, including those which occur in nature.

BOX 6.1
Temperature, terrapins, water fleas and micro-greenhouses

Heating and air conditioning systems insulate many of us from the vagaries of the climate but for most living organisms the temperature of their environment can have an important influence on their size and, hence, on a whole range of other biological functions: metabolic rates, fecundity, longevity and ability to migrate, compete, and withstand desiccation and starvation. The effects are especially important for plants and for ectotherms — reptiles, fish, insects, and so on, that rely on external sources for body heat and represent the vast majority of animals. Furthermore, the prospect of global warming has significant implications. Predicted increases in mean air temperatures of between 1.5 °C and 4.5 °C following increases in atmospheric CO_2 levels would result, for instance, in marked changes in the populations and distributions of many insect species.

Biological processes generally involve complex sequences of chemical steps, yet in common with many other

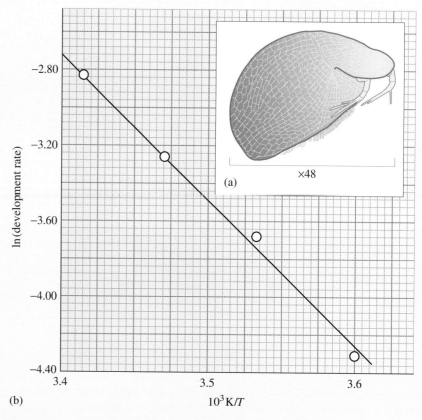

Figure 6.6 (a, inset) The water flea (*Alona affinis*) and (b) an Arrhenius plot of the rate of its development.

composite reactions, they often exhibit an Arrhenius-type behaviour. As an example, the rate of development, from egg-laying to adult, of a species of water flea (*Alona affinis*) taken from the River Thames near Reading (Figure 6.6a), appears to have an 'activation energy' of 65 kJ mol^{-1} (Figure 6.6b). By contrast, the heartbeat of a terrapin between 18 °C and 34 °C is driven by a process with an 'activation energy' of 73 kJ mol^{-1}, whereas at a lower temperature some other method of control takes over with a much higher activation energy (Figure 6.7). Similarly, a broken Arrhenius plot, indicative of different mechanisms in different temperature ranges, is found for the speed of creeping of a species of ant *Liometopum apiculetum*. It is also worth mentioning that the chirping rate of Mediterranean cicadas is so well related to ambient temperature that it can be used to 'hear' the temperature.

(a)

Temperature conditions are also important over long periods of time for the evolutionary progress of a given species, an example being seen in the flower shapes adopted by many wild plants in the Arctic (Figure 6.8). By arranging the petals to trap solar energy, the temperature is increased at the centre of the flower, speeding up the biochemical reactions in the reproductive organs and allowing seed formation to be completed during the short summer season. The bell-shaped, rather pendulous, flowers of heathers collect warm air rising from the ground, while the bowl-shaped flowers of the arctic poppy track the sun continuously and in the absence of wind and cloud can attain temperatures 10 °C above ambient. The result is an increase in both the number and weight of seeds reaching maturity.

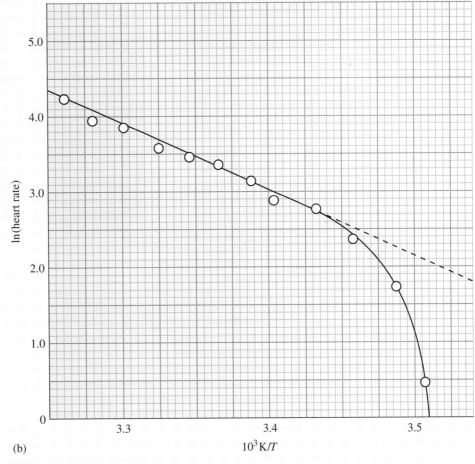
(b)

Figure 6.7
(a) A diamond-backed terrapin and (b) an Arrhenius plot of its heartbeat (rate) in the temperature range 18 °C to 34 °C. At lower temperatures, the plot departs from linear behaviour indicating a different control method.

(a)   (b)

Figure 6.8 Wild plants in the Arctic: (a) a heather (*Cassiope tetragona*) and (b) the Arctic poppy (*Papaver radicatum*).

Arctic strategies are obviously less important for temperate plants but they can still be advantageous, especially for species such as the garden *Crocus*, some varieties of which flower in Britain between October and April when ambient temperatures are low and the sunlight may be weak (Figure 6.9).

Figure 6.9 Because of their shape, crocus flowers act as micro-greenhouses generating internal temperatures which can be 3 °C above that of the surrounding air.

6.4 Summary of Section 6

1   The experimental rate constants for a large number of chemical reactions depend on temperature in a way that can be described by the Arrhenius equation

$$k_R = A \exp\left(-\frac{E_a}{RT}\right)$$

2   The Arrhenius parameters are the activation energy, E_a, and the A-factor, A. The units of E_a are typically $kJ\,mol^{-1}$ and those of A are the same as the units for the rate constant.

3   The value of the activation energy can be obtained from a plot of $\ln k_R$ (or $\log k_R$) versus $1/T$: this is referred to as an Arrhenius plot. The slope of the resulting straight line is equal to $-E_a/R$ (or $-E_a/2.303R$ if logs to the base 10 are used) so that E_a can be determined. Any point on the line may then be used to determine the A-factor. Alternatively, the A-factor can be determined from the intercept of the line (which is equal to $\ln A$) as calculated using a suitable fitting procedure.

4   At a given temperature, the rate of a reaction will depend upon the value of the *A*-factor, the magnitude of the activation energy, and the concentration(s) of the reactant(s). For many reactions it is the magnitude of the activation energy that is most influential. As a rough generalization, rapid reactions are characterized by relatively small activation energies.

5   If we consider a reaction at two different temperatures $T(1)$ and $T(2)$, then the rate constants at these temperatures, $k_R(1)$ and $k_R(2)$, are related as follows

$$\ln\left(\frac{k_R(2)}{k_R(1)}\right) = \frac{E_a}{R}\left(\frac{1}{T(1)} - \frac{1}{T(2)}\right)$$

Calculation of the ratio $k_R(2)/k_R(1)$ requires only that the activation energy for the reaction be known. This ratio will also be equal to the ratio of the rates of reaction providing that the initial concentration conditions at the two temperatures are the same.

EXERCISE 6.1

Values of the first-order rate constant for the gas-phase decomposition of chloroethane to give ethene and hydrogen chloride

$$C_2H_5Cl(g) = C_2H_4(g) + HCl(g)$$

are given in Table 6.3 for reactions carried out in the temperature range 670 K to 800 K. (These high temperatures were required in order to obtain measurable rates.) Use the information in Table 6.3 to determine the Arrhenius parameters for this decomposition reaction.

Table 6.3   Variation of the first-order rate constant with temperature for the gas-phase decomposition of chloroethane

$\dfrac{T}{K}$	$\dfrac{k_R}{s^{-1}}$
673.8	1.19×10^{-5}
699.6	6.10×10^{-5}
717.6	1.65×10^{-4}
727.2	3.00×10^{-4}
745.8	8.46×10^{-4}
764.8	2.42×10^{-3}
779.2	4.16×10^{-3}
793.2	8.41×10^{-3}

ELEMENTARY REACTIONS

For the rest of Part 1 of this book, we turn our attention to the discussion of just how chemical reactions occur and, as a consequence, we adopt a more theoretical viewpoint. To begin such a discussion we extend the description of elementary reactions that we started in Section 2.1. You should recall from that section that an elementary reaction is one which takes place in a single step, does not involve the formation of any intermediate species, and which passes through a single transition state.

7.1 Molecularity and order

A common method of classifying elementary reactions uses the *number* of reactant particles that take part in the reaction; this is called the **molecularity**. Elementary reactions are thus classified as *unimolecular*, *bimolecular* or *termolecular* depending on whether one, two or three reactant particles are involved, respectively. In fact, there are very few examples of elementary reactions that are thought to be termolecular and higher molecularities are unknown; it would be highly improbable that four reactant species could all collide and undergo chemical transformation at the same instant in either a gas-phase or solution reaction.

The stoichiometries of unimolecular and bimolecular reactions can be expressed in the following general forms

$\quad$ *unimolecular*: A $\longrightarrow$ product(s) $\hfill$ (7.1)

$\quad$ *bimolecular*: A + B $\longrightarrow$ product(s) $\hfill$ (7.2)

where arrows have been used to indicate that the reactions are elementary. As shown in Section 4.1, a simple collision model can be used to predict that the *theoretical* rate equation for an elementary reaction of the form of Reaction 7.2 is

$\quad J = k_{\text{theory}}[A][B] \hfill$ (4.6)

This prediction is also borne out experimentally (cf. Table 4.1) since the *experimental* rate equation is of exactly the same form; in other words the partial order of reaction with respect to each reactant is 1 and overall the reaction is second-order. In the case of a unimolecular reaction, the situation is more complicated since a single reactant particle has to become energized or *activated* by collisions, either with other reactant particles or other bodies that are present, in order for reaction to occur. However, although we shall not go into detail, a theoretical treatment shows that under most circumstances the rate of reaction will be directly proportional to the concentration of the single reactant species so that the theoretical rate equation can be written

$\quad J = k_{\text{theory}}[A] \hfill$ (7.3)

Once again this prediction is borne out by experiment so that the partial order with respect to the single reactant is 1 and overall the reaction is first-order.

A special case of a bimolecular reaction is that involving identical reactant species; for example, an atom recombination reaction such as

$\quad 2Br \longrightarrow Br_2 \hfill$ (7.4)

or written more explicitly

$$Br + Br \longrightarrow Br_2 \qquad (7.5)$$

The derivation of the form of the theoretical rate equation using a simple collision model is more complex in this type of case but it turns out that the equation is of the expected form

$$J = k_{theory}[Br]^2 \qquad (7.6)$$

and this is in agreement with experiment.

The discussion, so far, highlights a very important feature of elementary reactions: *the experimental overall order of reaction is the same as the molecularity*. In turn this means that if a reaction is proposed to be elementary then, on the basis of this information alone, the form that the experimental rate equation will take can be written down.

QUESTION 7.1

The following reactions are thought to be elementary. In each case write down the form of the *experimental* rate equation.

(a) The dissociation of molecular iodine into iodine atoms

$$I_2 \longrightarrow I + I$$

(b) The reaction between a hydrogen molecule and an iodine atom

$$H_2 + I \longrightarrow HI + H$$

(c) The isomerization of methyl isonitrile

$$CH_3NC \longrightarrow CH_3CN$$

It is important to recognize that molecularity is a theoretical concept that only refers to a reaction that is thought to be elementary; it has a value that is usually 1 or 2. The overall order of a reaction is a quantity that is determined *experimentally* and, depending on the form of the rate equation, can apply equally well to reactions that are elementary or composite.

⬤ If a reaction is first-order, what can be said about its molecularity?

⬤ A first-order reaction could be elementary or composite and since no information is provided to distinguish between the two it is not possible to say whether it is unimolecular, or not.

7.2 Reactions in the gas phase

In Section 4.1 we discussed a simple collision model to describe an elementary reaction in the gas phase between two reactant species A and B, in other words a bimolecular reaction.

⬤ Can you recall the essential features of this simple collision model?

⬤ There were two key features. First, for a reaction to occur there must be a collision between the reactant species and, second, the energy involved in this collision must be sufficient to overcome the energy barrier to reaction separating the reactants from the products.

In the early 1920s these ideas were developed in some detail. In particular, attention was focused on calculating the value of the second-order rate constant for a bimolecular reaction in the gas phase. It is important to recognize that experimentally such a rate constant will reflect in some way an average over many billions of individual reactions. It was therefore essential to develop a model of behaviour at the molecular level that could be averaged in a suitable way in order to make comparisons with experiment. Effectively, this restricted the theoretical development to gases since the behaviour of particles in a gas was far better understood than the corresponding behaviour in a solution.

Daniel Bernoulli (1700–1782)

Daniel Bernoulli was born into a Swiss family of mathematicians and scientists. At the age of 23 he became Professor of Mathematics at St Petersburg Academy but returned to Basel in 1733 to become, initially, Professor of Anatomy, then Botany and finally Natural Philosophy. He had the distinction of gaining the French Academy Prize ten times. His Kinetic Theory of Gases put forward in 1738 was developed over 100 years later by James Clerk Maxwell and Ludwig Boltzmann. The theory considers molecules (or atoms) in an ideal gas to be a collection of minute hard spheres. Applying the Newtonian laws of motion, together with statistical methods to average over the huge number of particles in a normal gas sample, it gives quantitative relationships for many of the fundamental properties of a simple gas.

Figure 7.1
Daniel Bernoulli (1700–1782).

In the **collision theory** reactant species are represented as hard spheres with definite radii, similar to snooker balls. Thus, the theory abandons any attempt to take into account the chemical structure of reactants. Furthermore, all attractive forces between species are ignored and a very large repulsive force is assumed to exist between them when they collide; that is, the hard spheres do not deform in any way. Figure 7.2 illustrates a typical collision. The assumptions are no doubt drastic but have the advantage that the mathematics becomes tractable.

To develop the theory it is necessary to take into account the fact that the reactant species A and B will each have a characteristic distribution of molecular speeds as given by the Maxwell distribution (see Figure 4.1). The distribution not only influences the number of collisions that occur between A and B at a given temperature, but it also plays a role in determining the fraction of these that can lead to reaction. In essence, the collision theory introduces a **threshold energy**, E_0, for reaction. The significance of this is that it is only those collisions between A and B that have a combined kinetic energy at impact greater than E_0 that can result in reaction. The final result of all of the mathematics, although we shall not delve into the detail here, is an expression for the bimolecular second-order rate constant, k_{theory}, which takes the form

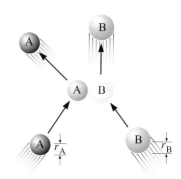

Figure 7.2
A schematic view of a collision between two reactant species A and B. Each species is represented by a hard sphere with a particular radius, that is r_A and r_B, respectively.

$$k_{theory} = A_{theory} \exp\left(-\frac{E_0}{RT}\right)$$

(7.7)

● Do you recognize the form of this equation?

● It is similar in form to the Arrhenius equation (Equation 6.3), $k_R = A \exp(-E_a/RT)$

Thus, at least qualitatively, a hard-sphere collision model results in an equation which has a firm basis in experiment. This suggests that the Arrhenius parameters, which were treated as purely *experimental* parameters in Section 6, can be given a physical interpretation, *at least for bimolecular processes.*

The theoretical A-factor

According to Equation 6.3, this factor is equivalent to the Arrhenius A-factor. In the collision model it is a measure of the standard rate at which reactant species collide; that is it is a measure of the number of collisions per second when the concentrations of the reactant species are both 1 mol dm^{-3}. It is necessary to specify standard conditions since, in general, the collision rate depends on the concentrations of the species present (cf. Section 4.1). The value of A_{theory} for a given bimolecular reaction depends on the hard-sphere radii and masses of the reactant species. Calculations show that it does not vary significantly from reaction to reaction with values usually of the order of 10^{11} dm^3 mol^{-1} s^{-1}. Table 7.1 compares the calculated values of A_{theory} for a few gas-phase bimolecular reactions with those derived from experiment.

Table 7.1 A comparison of experiment with collision theory for a selection of gas-phase bimolecular reactions

Reaction	$\dfrac{A_{\text{experiment}}}{\text{dm}^3\ \text{mol}^{-1}\ \text{s}^{-1}}$	$\dfrac{A_{\text{theory}}}{\text{dm}^3\ \text{mol}^{-1}\ \text{s}^{-1}}$
$H_2 + Cl \longrightarrow HCl + H$	8.3×10^{10}	3.6×10^{11}
$NO_2 + O_3 \longrightarrow NO_3 + O_2$	6.3×10^9	1.7×10^{11}
$H_2 + C_2H_4 \longrightarrow C_2H_6$	1.2×10^6	7.3×10^{11}

The comparison in Table 7.1 is typical in that there is not always good agreement between experiment and calculation with a tendency for the experimental values of the A-factor to be smaller, often markedly so, than the calculated values. Onc source of the problem undoubtedly lies in the treatment of the reactants as featureless spheres, which means that the theory takes no account of the relative orientation of the colliding particles. For any species more complicated than a free atom, however, orientation may well be important. For example, if the point of impact of one species does not occur at a reactive site in another then the probability of a successful outcome decreases. Figure 7.3 illustrates this idea in a schematic manner for the gas-phase reaction between potassium atoms and iodomethane molecules

$$K + CH_3I = CH_3 + KI \tag{7.8}$$

Detailed experiments in which a stream of potassium atoms was fired at a stream of iodomethane molecules, a so-called molecular beam experiment, revealed that the situation in which a potassium atom collided with the iodine end of the iodomethane molecule, approach (b) in Figure 7.3, was more likely to lead to reaction.

Overall it seems unlikely that it is just orientational or geometric effects that can fully account for the discrepancies between the collision theory and experiment. Nonetheless, although fairly rudimentary, the theory does provide a useful means of picturing how reactions occur.

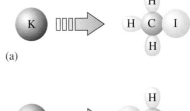
(a)

(b)

Figure 7.3
A schematic view (not to scale) of two possible collision approaches (a) and (b) for a potassium atom and an iodomethane molecule.

There is one final point to make about the value of A_{theory}. Calculation predicts that it should depend on the square root of temperature, that is $\sqrt{T}$. If such a dependence is also applicable to the experimental A-factor, this dependence will only have a small effect as the temperature is raised. For example, for a reaction with an activation energy of $E_a = 100\,kJ\,mol^{-1}$, a temperature rise from 300 K to 330 K will cause the term $\exp(-E_a/RT)$ to increase by a factor of almost 40, but the A-factor will increase by only a factor of 1.05. Thus, except over wide temperature intervals, it is reasonable to assume that the Arrhenius A-factor is independent of temperature and that the variation of reaction rate with temperature is controlled by the magnitude of the activation energy (cf. Section 6.3).

The threshold energy, E_0

Collision theory provides no means of calculating this energy; it is simply introduced into the model as a parameter, albeit a very important one. As already discussed, it represents the energy above which collisions will lead to successful reaction. Comparing the Arrhenius equation with Equation 7.7, thus suggests a similar physical interpretation for the activation energy; effectively it represents an *energy barrier to reaction* for a bimolecular process. In fact Arrhenius foresaw this physical interpretation in his original work (1889) on the inversion of cane sugar by acids:

> '… The actual reacting substance is not cane sugar… but is another hypothetical substance … which we call active cane sugar … formed from cane sugar at the expense of heat …'

Thus, before the cane sugar could react, it had to become 'activated' by gaining an amount of energy at least equal to the activation energy E_a needed to surmount the energy barrier to reaction.

In Section 2.1 we introduced the idea of drawing an energy profile for an elementary reaction. Sketch such a profile for an exothermic bimolecular reaction and indicate on the profile the activation energy for the reaction.

The energy profile is shown in Figure 7.4 and the activation energy, E_a, is shown as the energy difference between the top of the energy barrier and the level of the reactants. Since the activation energy represents a quantity that is measured experimentally, it is best to view the energy profile as representing some form of average over many individual bimolecular reactions.

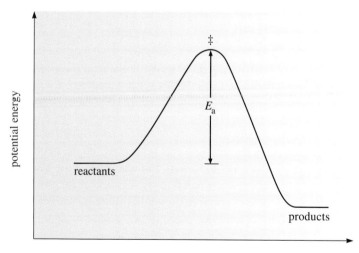

Figure 7.4 An energy profile for an exothermic bimolecular reaction indicating the activation energy E_a.

Since the Arrhenius equation applies equally well to composite reactions, it is sometimes the case that an energy profile such as that in Figure 7.4 is also used to represent these reactions. Strictly this is not correct, although it can be useful as a shorthand representation. A composite reaction consists of a sequence of elementary steps each of which will have its own energy profile and activation energy. The activation energy that is derived from an experimental Arrhenius plot for a composite reaction will represent some combination of the activation energies for the individual steps. It should be noted that the concept of an activated complex, whose presence is indicated on the energy profile in Figure 7.4 by the ‡ symbol, is an integral component of a more sophisticated theory of elementary reactions. We do not consider this theory here.

7.3 Reactions in solution

In large part, the chemistry we meet in practice takes place in a solution of some kind, but a quantitative description of the chemical kinetics involved is much more complex than for gaseous reactions. The key difference lies in the interparticle distances. In a gas at atmospheric pressure, the particles occupy less then 1% of the total volume and, effectively, move independently of each other. In a solution the solute and solvent molecules, with the latter being in the majority, take up more than 50% of the available space, the distances between the various species are relatively small, and each particle is in continuous contact with its neighbours. It is these interactions which greatly complicate the formulation of a satisfactory theory of chemical kinetics in solution. Indeed, the rate of an elementary reaction and for that matter a composite reaction, can be significantly influenced by the choice of solvent.

The simplest description of the rate of a bimolecular reaction in solution, equivalent to the collision theory in the gas phase, is based on the fact that a reactant species in a dilute solution is surrounded rather closely by a cage of solvent molecules: a **solvent cage** (Figure 7.5a). This cage is not static since solvent molecules are continuously entering and leaving it; however, on average, the number of molecules making up the cage remains the same. A reactant species can thus be envisaged as diffusing through a solution in a series of discontinuous jumps as it squeezes between the particles in the wall of one solvent cage and breaks through into a neighbouring cage. These jumps will be relatively infrequent and so reactant species meet much less often than they would in the gas phase. However, when they do find themselves in the same cage (Figure 7.5b) they remain there for a relatively long time. This period is known as an **encounter**. During it, the reactant species, which together are known as the **encounter pair**, retain their separate identities and may undergo a large number of collisions (with each other as well as with the cage walls) until eventually they either react or escape to separate cages (Figure 7.5c–e).

Figure 7.5
(a) Separate cages around two different reactant species in a dilute solution; (b) diffusion of the reactant species into the same cage at the beginning of an encounter; (c) a reactant–reactant collision during the encounter; (d) reactant-solvent collisions during the encounter; (e) break-up of the encounter pair with no reaction having occurred.

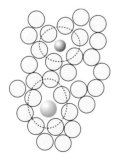

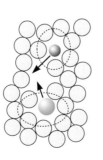

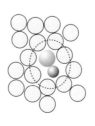

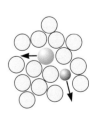

(a)　　　　(b)　　　　(c)　　　　(d)　　　　(e)

Two essential differences can be identified between a bimolecular gas-phase and a solution-phase reaction. They are

- Collisions between reactant species in the gas phase take place at fairly regular intervals, whereas in solution they occur in intermittent bunches.

- Successful reaction in the gas phase requires the two reactant species to have an energy exceeding a certain threshold at the instant of collision. This is not necessary at the beginning of an encounter in solution; reactant species entering a cage with insufficient energy may react later in the encounter by accumulating additional energy by way of the collisions they undergo with the cage walls.

○ Can you suggest what will happen in an encounter if the threshold energy for reaction between the two species is very low?

○ Once the encounter pair forms, then reaction is assured, probably at the first collision.

Thus the rate of reaction for a bimolecular reaction with a low threshold energy will be determined by the rate at which encounter pairs form. In turn this will depend on how rapidly the reactant species can diffuse through the solution and this is directly related to viscosity; the more viscous the solution, the more difficult it is for the reactant species to diffuse on a microscopic scale. In these circumstances, the reaction is said to be under **diffusion control**. The signature of such reactions is that the experimental activation energy is relatively small, lying typically in the range $10\,kJ\,mol^{-1}$ to $25\,kJ\,mol^{-1}$.

○ If an activation energy significantly greater than $25\,kJ\,mol^{-1}$ is measured for a bimolecular reaction in solution, can you suggest how to interpret this particular result?

○ The activation energy must more closely reflect the threshold energy for the *chemical* reaction between the reactant species. Once an encounter pair has been formed, a large number of collisions, including those with the walls of the cage, must take place before reaction can occur.

Bimolecular reactions in solution with activation energies significantly greater than $25\,kJ\,mol^{-1}$ are said to be under **activation control**.

7.4 Femtochemistry

Collision theory, as we have already seen, provides no direct information on exactly what happens when individual reactant molecules reach the top of the energy barrier to reaction. A theoretical picture at this molecular level eventually emerged in the 1930s with the idea of an activated complex (Section 2.1) and it was immediately clear that the lifetime of such a species must be extremely brief. We can gain a rough estimate of how brief by making two assumptions. First, that a complex will have disintegrated into products when one of its bonds has stretched from an initial length typical of that for a normal chemical bond (for simplicity say 0.1 nm) to a length around 0.3 nm. Second, that in breaking this bond the two fragments of the complex move apart at a speed of, say, 1 km s^{-1}; which is similar to the mean speeds of normal particles in the gas phase.

⬤ What is the lifetime?

⬤ The lifetime will be the time taken to travel the distance stretched by the bond, that is 0.2 nm, at a speed of 1 km s⁻¹. This will be

$$\text{lifetime} = \frac{0.2 \text{ nm}}{1 \text{ km s}^{-1}}$$
$$= \frac{0.2 \times 10^{-9} \text{ m}}{1 \times 10^{3} \text{ m s}^{-1}}$$
$$= 2.0 \times 10^{-13} \text{ s}$$

This is equivalent to 200 femtoseconds where 1 femtosecond (1 fs) is equivalent to 10^{-15} s.

The result of the calculation, although approximate, means that it would be necessary to make measurements every few femtoseconds in order to directly follow the passage of an activated complex through the transition region. Such a project in the 1930s lay far beyond the bounds of experimental possibility. Some idea of the enormity of the problem can be gained, as one author puts it, by realizing that a femtosecond is to one second, as a second is to 32 million years! By the 1960s, techniques had been developed that were fast enough to observe short-lived *reaction intermediates* with lifetimes of 10^{-3} s to 10^{-6} s. Two decades later, the use of equipment that allowed beams of reactant species to be fired at one another (that is, crossed-molecular beam techniques) reduced observation times to the picosecond range (1 ps = 1×10^{-12} s). Today, the femtosecond goal has finally been achieved through a combination of molecular-beam and pulsed-laser technologies, work for which Ahmed H. Zewail received the 1999 Nobel prize in Chemistry.

Ahmed H. Zewail (1946–)

Ahmed H. Zewail was born in 1946 near Alexandria in Egypt and undertook undergraduate studies at the University of Alexandria. His Ph.D. was obtained from the University of Pennsylvania. Currently, he is the Linus Pauling Professor of Chemistry and Professor of Physics, and the Director of the National Science Foundation Laboratory for Molecular Sciences, at the California Institute of Technology (Caltech) in Pasadena. He has been awarded many honours including the 1999 Nobel prize in Chemistry. From Egypt he received the Grand Collar of the Nile, the highest state honour, and postage stamps were issued to celebrate his contributions to both science and humanity.

Figure 7.6   Ahmed H. Zewail (1946–).

In femtosecond spectroscopy (Figures 7.7 and 7.8), an ultra-fast laser injects two pulses of radiation into a beam of reactant molecules. In the case of the investigation of a unimolecular reaction these are all of one type. For a bimolecular reaction the situation is more complex but techniques exist to make such studies possible. The basics of the experiment are reasonably straightforward. The first laser flash, called the *pump pulse*, has photons of energy sufficient to be absorbed by individual reactant species and so generate activated complexes. This is taken as the starting point of the reaction and the activated complexes are at the top of the energy barrier at this point. A second, weaker laser flash, called the *probe pulse*, arrives a short time later (measured in femtoseconds) and this has a wavelength selected to excite a spectrum for the activated complex. A series of such spectra, obtained by varying the delay between the two pulses, essentially provides a series of snapshots of what is happening as the activated complex passes through the transition state and disintegrates to form the products.

Figure 7.7 Part of the laser system for femtochemisty in Ahmed Zewail's laboratory which is referred to as 'femtoland'.

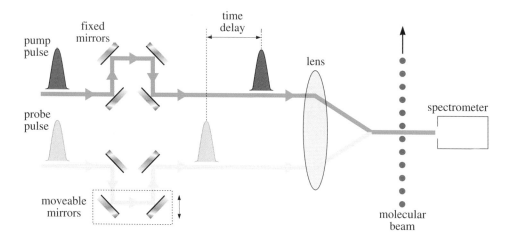

Figure 7.8 A schematic illustration of the experimental arrangement for femtochemistry. The pump pulse passes through a series of fixed mirrors, whereas the probe pulse can be delayed by passing through a series of movable mirrors.

Three of the factors that are crucial to the success of femtochemistry experiments are as follows. First, each laser pulse must last no longer than a few femtoseconds, otherwise there would be unacceptable uncertainty in identifying the moment at which a reaction is started or the time at which a spectrum is recorded. Second, perhaps obviously, the probe pulse must reach the sample within a few tens of femtoseconds after the arrival of the pump pulse. This is achieved by using mirrors to create a detour in the path of each laser beam, with the detour being longer for the probe pulse than for the pump pulse.

● Given that the speed of light is $3 \times 10^8 \, \text{m s}^{-1}$, to what distance does a delay of, say, 33 femtoseconds correspond?

● The distance will be given by speed × time, that is

$$distance = (3 \times 10^8 \, \text{m s}^{-1}) \times (33 \times 10^{-15} \, \text{s})$$
$$= 9.9 \times 10^{-6} \, \text{m}$$
$$= 9.9 \times 10^{-3} \, \text{mm}$$

So, the distance is roughly 0.01 mm.

Although relatively small, path differences of this order can be obtained and varied readily by adjustment of the movable mirrors in the path of the probe pulse (Figure 7.8).

Finally, since the spectrum of an activated complex in the transition state will be unknown in advance, setting up the experiment and interpreting the experimental data will rely on the ability to calculate spectra for a range of hypothetical activated complexes with different possible structures. The use of computer modelling methods is vital to this process.

The first direct observations within the transition state were achieved by Zewail and co-workers in 1987. The reaction investigated was the unimolecular decomposition of iodocyanide

$$ICN \longrightarrow I + CN \tag{7.9}$$

and they managed to observe the breaking of the iodine–carbon bond in the activated complex; a process that took about 200 fs. One of the first bimolecular reactions to be studied was one that is found as a step in combustion reactions, as well as in the chemistry of the atmosphere

$$H + CO_2 \longrightarrow HO + CO \tag{7.10}$$

The hydrogen atom was found to attack the carbon dioxide molecule and stick to it for hundreds of femtoseconds before stripping off an oxygen atom from the carbon. The OH species finally emerges 5 picoseconds after the start of the reaction.

An interesting question arises in relation to decomposition reactions in which two identical chemical bonds are broken. For example, in the formation of tetrafluoroethene (C_2F_4) from tetrafluorodiodoethane ($C_2F_4I_2$), that is

$$\tag{7.11}$$

it is reasonable to ask whether the carbon–iodine bonds break simultaneously or sequentially. The answer, in this instance at least, is the latter. Femtochemistry finds that one bond breaks some 200 fs after activation of the reactant and then cleavage of the second bond takes place after 20 ps, that is a length of time greater by a factor of 100.

Femtochemistry also has potential biological applications. For example, photochemical or light-induced processes are important in various areas of biology and physiology. Zewail's group has observed that excitation of the cis-stilbene molecule weakens the double bond and allows a *synchronous* rotation of each benzene ring through 90° in opposite directions to give the *trans*-isomer (Figure 7.9).

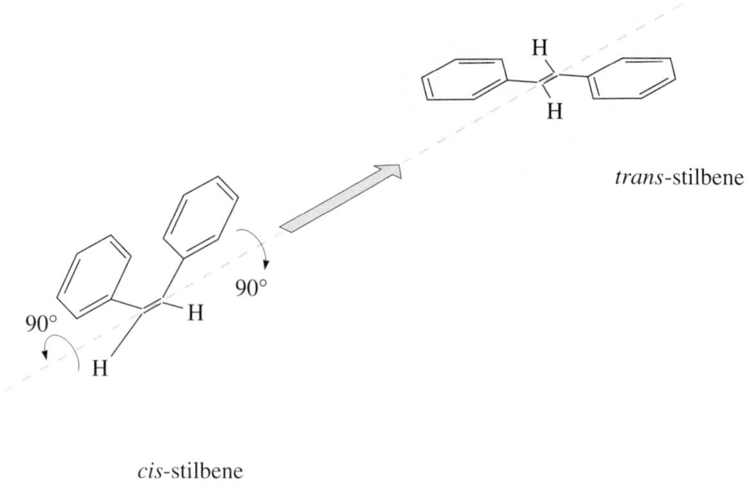

Figure 7.9 The synchronous rotation of each benzene ring through 90° to convert a *cis*-stilbene molecule to its *trans*-stilbene isomer.

In the visual process in the eye, a key photochemical step involves a similar *cis* to *trans* conversion of a double bond in an unsaturated aldehyde which is combined with other substances in the red pigment called rhodopsin in the rod cells of the retina. Other workers using femtosecond spectroscopy have found this process to be rapid, taking about 200 fs, which suggests that when a photon is absorbed, its energy is not distributed throughout the aldehyde molecule but rather it is localized directly in the relevant double bond. This would explain the high (70%) efficiency of the light-induced isomerization step and hence the eye's good night-vision. It has been suggested that there is a similarly efficient energy conversion in the chlorophyll molecules that capture light during photosynthesis in plants.

Femtochemistry has enabled chemical kinetics to progress from descriptions in terms of rather vague concepts such as activation and transition state to a much more detailed picture of molecules in the act of crossing energy barriers. Nevertheless, it does not yet mark the end of the road on which Wilhelmy, Arrhenius and others set out in the nineteenth century. Developments in laser technology will eventually lead to shorter pulses lasting not femtoseconds, but attoseconds (1 as = 1×10^{-18} s). Thus even more powerful techniques will become available to probe what actually happens in the act of chemical transformation as chemical bonds are broken and new ones are formed.

7.5 Summary of Section 7

1   The concept of molecularity provides a method of classifying elementary reactions in terms of the number of reactant particles that take part in the reaction.

2   Elementary reactions may be unimolecular, bimolecular or, very rarely, termolecular. An important feature of these reactions is that the experimental overall order of reaction is the same as the molecularity. This means that the theoretical rate equation for a reaction that is proposed to be elementary can be written down on the basis of the stoichiometry.

3   Molecularity is a theoretical concept and it should not be confused with reaction order which is an experimentally determined quantity.

4   Collision theory for a bimolecular reaction in the gas phase treats the individual reactant species as hard spheres and introduces a threshold energy for the reaction. The expression derived for the temperature dependence of the bimolecular second-order rate constant is of the same form as that for the Arrhenius equation. The theoretical A-factor is related to the rate at which reactant species collide and is calculated to be of the order of $10^{11}\,dm^3\,mol^{-1}\,s^{-1}$, although experimental values can be smaller than this by several orders of magnitude. The threshold energy, E_0, which is introduced as a parameter in the theory, provides a theoretical interpretation for the Arrhenius activation energy, E_a.

5   The simplest model for a bimolecular reaction in solution envisages reactant species surrounded by a cage of solvent molecules. An encounter occurs when two reactant species share the same cage; they are referred to as an encounter pair. Reaction can occur immediately on an encounter in which case the reaction is said to be under diffusion control (E_a typically lies in the range $10\,kJ\,mol^{-1}$ to $25\,kJ\,mol^{-1}$). Alternatively, it may take many collisions during the encounter, including those with the cage wall, before reaction occurs in which case the reaction is said to be under activation control (E_a significantly greater than $25\,kJ\,mol^{-1}$).

6   Femtochemistry is able to investigate what happens during the transition state of an elementary reaction on a timescale of femtoseconds ($10^{-15}\,s$). It involves a combination of molecular-beam and pulsed-laser technologies, and advanced theoretical calculations. It can, for example, provide information on whether identical chemical bonds in a molecule are broken simultaneously or sequentially during a reaction.

REACTION MECHANISM

8.1 Evidence that a reaction is composite

As described in Section 2, a reaction mechanism refers to a molecular description of how reactants are converted into products during a chemical reaction. In particular, it refers to the sequence of one or more elementary steps that define the route between reactants and products. A prime objective of many kinetic investigations is the determination of reaction mechanism since this is not only chemically interesting in its own right, but it may also suggest ways of changing the conditions to make a reaction more efficient.

There is no definite prescription for working out a mechanism and in any particular situation a solution to the problem will depend on information from a variety of sources. The behaviour of similar systems, if already established, will be an important guide and, of course, any proposed mechanism must be reasonable in terms of general chemical principles. For example, a proposed reaction intermediate, even if it cannot be detected experimentally, must still be chemically possible.

One of the most important questions when dealing with reaction mechanism is whether or not the mechanism is composite and various kinds of evidence can often show clearly that a reaction does not occur in a single elementary step. Some of the most important forms of evidence are described below.

Detection of reaction intermediates

Any reaction that occurs via a series of steps will involve intermediate species; indeed the demonstration of the presence of just one reaction intermediate implies that a reaction must involve at least two steps and so must be composite. In some instances the stability of an intermediate can be sufficient to allow it to be isolated and characterized. However, more commonly, intermediates are highly reactive species and so exist in very low concentrations. If they can be detected at all, it is generally only by physical means such as some form of spectroscopy. Although the detection of an intermediate provides definite evidence for a composite reaction, it can be very difficult to achieve practically.

The form of the experimental rate equation

The most commonly used method of identifying a composite reaction is via an examination of the experimental rate equation. As discussed in Section 7.1, the experimental overall order of an elementary reaction is the same as its molecularity. In other words, the form of the experimental rate equation can be written down on the basis of *just* the reaction stoichiometry.

⬤ The decomposition of ozone (O_3)

$$O_3(g) = O_2(g) + O(g) \tag{8.1}$$

is thought to be an elementary reaction. If this is the case what is the expected form of the experimental rate equation?

⬤ The elementary reaction is unimolecular, so the experimental rate equation will be

$$J = k_R[O_3] \tag{8.2}$$

⬤ The reaction between nitrogen dioxide and ozone

$$NO_2(g) + O_3(g) = NO_3(g) + O_2(g) \tag{8.3}$$

is also thought to be elementary. What is the expected form of the experimental rate equation?

⬤ The reaction is bimolecular, so the experimental rate equation will be

$$J = k_R[NO_2][O_3] \tag{8.4}$$

The corollary of the above discussion is that if the experimental rate equation is different from that expected from the stoichiometry then the reaction *must be composite*. For many reactions it is just the complexity of their experimental rate equations that provides the required evidence. For example, the reaction between hypochlorite ion and iodide ion in aqueous solution

$$ClO^-(aq) + I^-(aq) = Cl^-(aq) + IO^-(aq) \tag{8.5}$$

has an experimental rate equation of the form

$$J = k_R \frac{[ClO^-][I^-]}{[OH^-]} \tag{8.6}$$

which involves the presence of a species, OH^-, which is not in the stoichiometric equation. There is no doubt that this reaction must be composite. Similarly, the reduction of the copper(II) ion (Cu^{2+}) by hydrogen gas in an aqueous acid solution

$$2Cu^{2+}(aq) + H_2(g) = 2Cu^+(aq) + 2H^+(aq) \tag{8.7}$$

has a very complex experimental rate equation:

$$J = \frac{k[Cu^{2+}]^2[H_2]}{k'[Cu^{2+}] + k''[H^+]} \tag{8.8}$$

where k, k' and k'' are all constants. The reaction must be composite.

⬤ If you were unaware of the form of the experimental rate equation for Reaction 8.7, can you suggest another reason which would make you suspect that the reaction was composite?

⬤ The reaction involves three reactant species ($Cu^{2+} + Cu^{2+} + H_2$) and so if it were elementary it would be termolecular. Termolecular reactions are very rare (Section 7.1) and so it would be reasonable to suspect that Reaction 8.7 was composite.

The experimental rate equations for some reactions have forms that are equivalent to those expected for unimolecular and bimolecular reactions but they *do not* correspond to the stoichiometry. For example, in Section 5.3.1 we established that the gas-phase thermal decomposition of dinitrogen pentoxide (Equation 5.1)

$$2N_2O_5(g) = 4NO_2(g) + O_2(g)$$

has an experimental rate equation of the form (Equation 5.2)

$$J = k_R[N_2O_5]$$

Thus, the experimental rate equation is first-order and so, if the decomposition were to be an elementary reaction, then the stoichiometry would have to correspond to that for a *unimolecular reaction*. This is clearly not the case since the stoichiometry involves *two* reactant molecules, albeit identical; indeed, the reaction stoichiometry could be written as

$$N_2O_5(g) + N_2O_5(g) = 4NO_2(g) + O_2(g) \tag{8.9}$$

The only conclusion to draw is that the reaction is composite.

A final note of caution

Our discussion has focused essentially on assembling evidence that a reaction cannot be elementary. However, there is another side to the argument which is less clear cut. Even if *none* of the tests already described (and other more specific ones that we have not mentioned) indicates a composite reaction, there is still no *absolute* guarantee that a reaction must be elementary.

> QUESTION 8.1
>
> The experimental rate equations for a series of reactions are given below. Identify, giving your reasons in each case, which of these reactions *must* have a composite mechanism.
>
> (a) The reaction between formate ion ($HCOO^-$) and permanganate ion (MnO_4^-) in aqueous solution
>
> $$3HCOO^-(aq) + 2MnO_4^-(aq) + 2H^+(aq) = 2MnO_2(s) + 3HCO_3^-(aq) + H_2O(l)$$
>
> which has an experimental rate equation of the form
>
> $$J = k_R[MnO_4^-][HCOO^-]$$
>
> (b) The reaction between hypochlorite ion and bromide ion (Equation 3.1 in Section 3)
>
> $$ClO^-(aq) + Br^-(aq) = BrO^-(aq) + Cl^-(aq)$$
>
> which has an experimental rate equation of the form (Section 5.4.1)
>
> $$J = k_R[ClO^-][Br^-]$$
>
> (c) The gas-phase thermal decomposition of ethanal (CH_3CHO)
>
> $$CH_3CHO(g) = CH_4(g) + CO(g)$$
>
> which has an experimental rate equation of the form
>
> $$J = k_R[CH_3CHO]^{3/2}$$

8.2 A procedure for simplification: rate-limiting steps and pre-equilibria

A proposed reaction mechanism should incorporate all of the information that is known about a particular reaction. *One key aspect is that it should be able to account for the form of the experimental rate equation.* Indeed, if the proposed mechanism cannot do this then it has to be discarded. Mathematically, the analysis of a reaction mechanism can be very difficult or even impracticable, and the complexity of the problem increases significantly with the number of proposed steps. Many of the problems, of course, have been eased with the advent of high-speed computing facilities but there are also various procedures available for simplifying the analysis of a mechanism. We consider one of these in this section. It is particularly important for the discussion of the mechanisms of substitution and elimination reactions in organic chemistry which are topics that are covered later in this book.

A possible *model* reaction mechanism for a reaction of the form

$$A = P \tag{8.10}$$

where A is a reactant and P is a product, can be written as

$$A \xrightarrow{k_1} X$$

$$X \xrightarrow{k_2} Y$$

$$Y \xrightarrow{k_3} P$$

In this mechanism the species X and Y are reaction intermediates and, overall, the mechanism consists of three consecutive elementary steps, each of which has its own rate constant. As a matter of notation, for a given step the rate constant is written above the arrow which is used to indicate that the reaction is elementary (Section 2.1); the arrow is elongated for this purpose. Each rate constant is given a subscript which identifies it as belonging to a particular step in the mechanism; so a subscript 1 indicates that the rate constant k_1 belongs to the first step in the mechanism, and so on. In this form the rate constant can be taken to represent either a theoretical or experimental value. Finally, as expected, for a composite mechanism the three steps add up to give the overall balanced chemical equation.

⬤ Write down the theoretical rate equation for each step in the mechanism. Label the rates of reaction as J_1, J_2 and J_3.

⬤ All three steps are unimolecular and so

Step 1: $J_1 = k_1[A]$

Step 2: $J_2 = k_2[X]$

Step 3: $J_3 = k_3[Y]$

The equations you have written down provide a mathematical description of the reaction mechanism. In fact, if we focus on the rate of production of the product P then all the equations are seen to be interlinked: the formation of P depends on the decomposition of Y, which in turn depends on the decomposition of X, which in turn depends on the decomposition of A. In principle, the equations can be solved to find how the concentrations of all of the species present (A, X, Y and P) vary with time but, overall, it is not possible to find a simple form of the rate equation. In fact, these comments need not be specific to the model mechanism we have chosen to discuss; in general, they will apply to any form of mechanism. However, we know from experimental investigations that many composite reactions which have time-independent stoichiometry have relatively simple rate equations. This must mean that it is legitimate to simplify the analysis of reaction mechanisms. But how? A clue to one approach can be taken from the unlikely source of an article in the *Daily News* (London) 26 December 1896:

> 'The widened portions (of the roadway) at Holloway and elsewhere are rendered useless by narrow, bottleneck approaches to Finsbury Park.'

In any multi-step process the different component reactions will proceed at different rates.

⬤ If one step were to be significantly slower than the others, what could you conclude about the overall rate of reaction as measured, say, by the rate of product formation?

It would be determined by the rate of the slow step, which would act as a bottleneck in the process.

The idea of a bottleneck in a reaction mechanism is an important one. In particular, for reactions with time-independent stoichiometry it can be used to simplify the analysis of a reaction mechanism and, in the process, derive a 'predicted' rate equation which can be compared with experiment. We shall not look at this in general, but we will consider specific cases.

To begin with, it is useful to return to Reaction 8.10 and specify that it has time-independent stoichiometry. In these circumstances the concentrations of the reaction intermediates X and Y will not build up to any significant extent during the course of the reaction; we can refer to them as *reactive* intermediates. If we now assume that the first step in the model reaction mechanism is slow then the second step will essentially have to wait on the formation of X before it can proceed. However, since X is a reactive intermediate once formed it will quickly react to give Y and, in turn, Y will quickly react to give the product P. The rates of these second and third steps may well be very rapid compared to the rate of the slow first step, but they will have *no* effect on the overall rate of reaction. The slow first step can be referred to as the **rate-limiting step** (sometimes the term *rate-determining* is used). This discussion leads to the following conclusions. For a reaction with time-independent stoichiometry and a reaction mechanism in which the first step is taken to be rate-limiting, the overall rate of reaction (J)

- will be the *same* as that of the rate-limiting step; and
- will *not* be influenced by the rates of any steps following the rate-limiting step.

As a more practical example, we can consider the conversion of hypochlorite ion to chlorate ion in aqueous solution which has the following time-independent stoichiometry

$$3ClO^-(aq) = ClO_3^-(aq) + 2Cl^-(aq) \qquad (8.11)$$

and a proposed reaction mechanism of the form

$$ClO^- + ClO^- \xrightarrow{k_1} ClO_2^- + Cl^-$$

$$ClO_2^- + ClO^- \xrightarrow{k_2} ClO_3^- + Cl^-$$

where the chlorite anion, ClO_2^-, is a reaction intermediate. (You may wish to check that the two steps in the mechanism add together to give the overall balanced chemical equation.)

 If the first step in the mechanism is rate-limiting, what is the predicted form of the experimental rate equation?

 The theoretical rate equation for the first step, which is bimolecular, is

$$J_1 = k_1[ClO^-]^2$$

Since this is also the rate-limiting step, the overall rate of reaction J will be equal to J_1 and so the form of the rate equation predicted from the mechanism will be

$$J = k_1[ClO^-]^2$$

The experimental rate equation for Reaction 8.11 is

$$J = k_R[ClO^-]^2 \qquad (8.12)$$

so there is good agreement with the rate equation predicted from the mechanism and the experimental rate constant (k_R) will be equal to the rate constant for the first step in the mechanism (k_1). This agreement between prediction and experiment does not provide *absolute* confirmation of the proposed mechanism but it does indicate that the mechanism is plausible.

Of course the analysis we have just discussed depends on the assumption that the first step is rate-limiting. In the particular case of the conversion of hypochlorite ion to chlorate ion it is possible to gain direct evidence since the chlorite ion (ClO_2^-) can be prepared separately and it can then be shown that its reaction with hypochlorite ion is fast. More generally, however, there are no comprehensive guidelines for identifying a rate-limiting process. In some cases, it is possible to eliminate certain steps. For example, any step known, or expected, to have a relatively *low* activation energy would *not* be a suitable candidate.

It should be clear that if the first step in a proposed reaction mechanism is taken to be rate-limiting then the analysis of the mechanism is very straightforward; the overall rate of reaction is simply equal to that of the first step and a comparison is made with the experimental rate equation. If this is successful, as with any other mechanistic investigation, it is then advisable to examine additional experimental information for corroboration of the key ideas. This can involve changing the reaction conditions or considering the relationship between the stereochemistry of the reactants and products. A much fuller discussion of the value of corroborative evidence is given in Part 2 which considers the mechanisms of organic substitution reactions.

The analysis of a proposed reaction mechanism in which the first step is not rate-limiting is a little more involved. As an example we shall consider a two-step reaction mechanism which is commonly proposed as a plausible mechanism for a wide variety of reactions.

Suppose the reaction

$$A + B = P \qquad (8.13)$$

has time-independent stoichiometry and a reaction mechanism consisting of two steps (Steps 1 and 2) of the form

Step 1: $\quad A \xrightarrow{k_1} X$

Step 2: $B + X \xrightarrow{k_2} P$

so that there is a single reaction intermediate X. If Step 2 is taken to be rate-limiting then this step will determine the overall rate of reaction.

⬤ In this case, what will be the form of the predicted rate equation?

⬤ The overall rate of reaction, J, will be equal to the rate of reaction of Step 2, that is

$$J = J_2 = k_2[B][X]$$

In these circumstances, the form of the expression for J is not very useful since it depends on the concentration of the reaction intermediate X. It is not possible to compare this expression with the experimental rate equation which will be expressed in terms of reactant concentrations. (In more complex cases, product concentrations may also appear in the experimental rate equation.) Indeed, from an experimental perspective the concentration of X, which is a reactive intermediate, may well be too small to be measured.

The situation may seem intractable. However, the fact that Step 2 is relatively slow must mean that the reaction intermediate X will be formed in Step 1 more rapidly than it can be consumed in Step 2 and so during the *initial* stages of the reaction its concentration will increase. However, it must be remembered that the reaction has time-independent stoichiometry and so this increase in the concentration of X must be moderated by another process. The key point is that as the concentration of X builds up, the rate of a new process, the *reverse* of Step 1, becomes significant. We represent this reverse process as

$$X \xrightarrow{k_{-1}} A$$

with the subscript attached to the rate constant being *negative* to indicate that it refers to the *reverse* of Step 1. The forward and reverse reactions of Step 1 must *both* be relatively rapid compared to the slow second step. Thus Step 1 will quickly reach a condition of equilibrium *within the reaction scheme* and this equilibrium will hardly be affected by the slow second step throughout the course of the reaction. In this context Step 1, because it *precedes* the rate-limiting step is referred to as a rapidly established **pre-equilibrium**.

The mechanism for the reaction is more usually written as

Step 1: $\quad A \underset{k_{-1}}{\overset{k_1}{\rightleftharpoons}} X$

Step 2: $B + X \xrightarrow{k_2} P$

(Note that summation of the steps in the mechanism to give the balanced chemical equation only holds if the reverse arrow is ignored in Step 1.) For Step 1 at equilibrium the rate of the forward and reverse reactions will be equal, that is

rate of forward reaction in Step 1 = rate of reverse reaction in Step 1 $\qquad$ (8.14)

What are the theoretical rate equations for the forward and reverse reactions in Step 1? (Label the rates of reaction as J_1 and J_{-1}, respectively.)

For the forward reaction

$$J_1 = k_1[A]$$

and for the reverse reaction

$$J_{-1} = k_{-1}[X]$$

If we equate these rates of reaction, as in Equation 8.14, then

$$k_1[A] = k_{-1}[X] \qquad (8.15)$$

We are now able to relate the concentration of X to a concentration that is *measurable*, that is the concentration of the reactant A; in fact rearranging Equation 8.15 gives

$$[X] = \frac{k_1}{k_{-1}}[A] \qquad (8.16)$$

Given that we have already expressed the overall rate of reaction as

$$J = k_2[B][X] \qquad (8.17)$$

then substituting directly for [X] gives

$$J = k_2[B]\left(\frac{k_1}{k_{-1}}[A]\right) \qquad (8.18)$$

This equation can be rearranged into a more familiar form by changing the order of the terms

$$J = \frac{k_2\,k_1}{k_{-1}}[A][B] \qquad (8.19)$$

Thus, with Step 2 rate-limiting and Step 1 a rapidly established pre-equilibrium, the proposed reaction mechanism *predicts* that the experimental rate equation will be second-order overall with the partial order of reaction with respect to each reactant equal to 1. Furthermore on comparison with an experimental rate equation of the form

$$J = k_R[A][B]$$

it can be seen that the experimental rate constant is predicted to be equal to

$$k_R = \frac{k_2\,k_1}{k_{-1}}$$

There is one further point to make about the analysis.

⬤ If the equilibrium constant for the first step in the reaction mechanism is represented by K_1, how will this equilibrium constant be related to the concentrations of A and X?

⬤ It will be related in the usual way

$$K_1 = \frac{[X]}{[A]}$$

The magnitude of K_1 must clearly be considerably less than unity reflecting the fact that the concentration of the reactive intermediate X is significantly less than that of the reactant A.

⬤ An alternative expression for the ratio [X]/[A] can be obtained from Equation 8.15. What is this?

⬤ If Equation 8.15 is rearranged then

$$\frac{[X]}{[A]} = \frac{k_1}{k_{-1}}$$

The equilibrium constant for the first step in the reaction mechanism can therefore be related to the rate constants for the forward and reverse reactions for this step, that is

$$K_1 = \frac{k_1}{k_{-1}} \qquad (8.20)$$

In turn, the predicted rate equation (Equation 8.19) can be written in the form

$$J = k_2K_1[A][B] \qquad (8.21)$$

and this is an equally valid representation that is widely used.

The discussion, so far, has focused on a particular form of two-step reaction mechanism: 'reactant A reacts to give a reactive intermediate X which then further reacts with reactant B to give product'. It is possible to envisage different variations on this theme. Nonetheless, if the reaction in question has time-independent stoichiometry and the second step in the reaction mechanism is taken to be rate-limiting then

• the second step will determine the overall rate of reaction (J), and

• the first step can be taken to be a rapidly established pre-equilibrium.

These two principles provide the basis on which the expression for a predicted rate equation can be derived. The following question which revisits the proposed mechanism for the conversion of hypochlorite ion to chlorate ion in aqueous solution (Reaction 8.11) provides an example.

QUESTION 8.2

Suppose that the second step in the proposed reaction mechanism for Reaction 8.11 had been suggested to be rate-limiting. What form would the predicted rate equation take in these circumstances?

As a final general point we should note that in many proposed mechanisms no single step stands out as being much slower than the rest. The procedure of simplifying the analysis of a reaction mechanism by assigning one step to be rate-limiting is therefore by no means universally applicable.

8.3 Confirmation of a mechanism

A prime requirement of the acceptability of a proposed reaction mechanism is that analysis of this mechanism gives rise to a predicted rate equation that is in agreement with experiment. Further supporting evidence can also be provided by additional experimental tests. A qualification, however, is necessary. As a description of how a reaction occurs, a mechanism is no more than a *hypothetical model*. It is a possibility, but not a certainty; there being no guarantee that it will be able to accommodate facts that come to light in the future (see, for example, Question 8.3). In the meantime, however, this does not make it any less useful than other types of theory or law in experimental science; there are none for which such a guarantee can be given.

QUESTION 8.3

In a series of papers published in the period 1894–98, M. Bodenstein showed that the gas-phase reaction between hydrogen and iodine

$$H_2(g) + I_2(g) = 2HI(g)$$

had a second-order experimental rate equation of the form

$$J = k_R[H_2][I_2]$$

For many years the reaction was thought to be a prime example of an elementary reaction. However, later work by J. H. Sullivan in 1967 (entitled, *Mechanism of the 'bimolecular' hydrogen–iodine reaction*) showed that iodine *atoms* were involved in the reaction so that it must be composite. A two-step reaction mechanism can be proposed, in which molecular iodine dissociates, and the iodine atoms then react with a hydrogen molecule

$$I_2 \longrightarrow I + I$$
$$2I + H_2 \longrightarrow 2HI$$

(The latter step in this mechanism, in fact, represents a combination of two steps. The first of these steps involves the formation of a weakly bonded complex, which is often represented as $H_2 \cdots I$, and the second step involves the interaction of this complex with another iodine atom. For the purposes of the question, it is acceptable to join these steps together to give an apparently termolecular reaction.)

Show that the two-step reaction mechanism can lead to a predicted rate equation which is in agreement with experiment.

BOX 8.1   Oscillations: combustion, heartbeats and arctic hares

For most of the reactions carried out in the laboratory, the concentrations of the participating species vary steadily as time progresses; those of reactants decrease, those of products increase and those of intermediates rise initially and then fall again. But this need not necessarily be so. For certain reactions under limited conditions, the concentrations of some of the species in the reaction mixture are found to oscillate, rising and falling repeatedly. Such reactions are called **oscillating reactions**.

Following a description in 1828 of an electrochemical cell which produced an oscillating current, a few examples of similar phenomena slowly came to light, such as a report in 1916 of the periodic liberation of carbon monoxide (CO) during a reaction caused by mixing methanoic (formic) acid (HCOOH) with concentrated sulfuric acid (Figure 8.1).

Figure 8.2
B. P. Belousov.

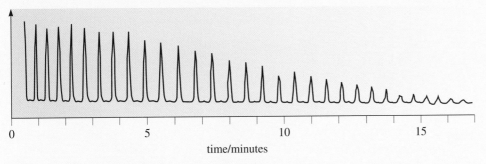

time/minutes

Figure 8.1   Periodic bursts of carbon monoxide observed when methanoic acid is dissolved in a concentrated solution of sulfuric acid at 55 °C. The vertical axis represents a rate of CO liberation measured in arbitrary units.

None of the cases discovered, however, was purely chemical in nature and all involved the physical process of diffusion. For example, in Figure 8.1 the bursts of carbon monoxide are due to tiny bubbles of the gas coming together to form larger bubbles which then diffuse to the surface and escape. It was believed, therefore, that in the absence of diffusion — in a *homogeneous* chemical system having the same concentrations throughout — oscillations could not occur and, indeed, at a deeper level of discussion they would be forbidden by the second law of thermodynamics. This belief persisted even though A. J. Lotka in 1910 devised a simple kinetic scheme which did predict such behaviour.

In the late 1950s, B. P. Belousov (Figure 8.2) when working in the Ministry of Health in Moscow made a chance observation. Working with a rather complicated but *well-stirred* mixture of reagents in which malonic acid, $CH_2(COOH)_2$, was converted to bromomalonic acid, $CHBr(COOH)_2$ in the presence of a cerium ion catalyst, he noticed that the concentration of an intermediate — the bromide ion Br^- — as well as the concentration of the cerium ion catalyst, rose and fell regularly. Convinced that these were genuine chemical oscillations, Belousov submitted the work for publication but it was dismissed as impossible.

Fortunately, a short report of the observations did appear in the proceedings of an obscure conference on radiation chemistry and in 1967 A. Zhabotinsky at Moscow State University was able to confirm the results. Less fortunately, Belousov died before he, Zhabotinsky and three others were awarded the 1980

Lenin prize. The B–Z reaction, as it is now known, has been studied intensively and many other cases of oscillations in homogeneous chemical systems have been identified. Figure 8.3 shows an example from the combustion of hydrogen. Similar fluctuations during the oxidation of hydrocarbon fuels are responsible for 'knock' in internal combustion engines.

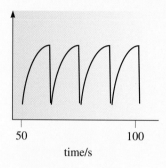

Figure 8.3
Oscillations in oxygen concentration when hydrogen and oxygen are reacted together in a flowing system at 700 K to 800 K. The vertical axis represents a measure of the oxygen concentration.

Oscillatory mechanisms tend to be complicated, involving about 20 steps in the case of the B–Z reaction. In general, to produce *sustained* oscillations, a reaction must be far removed from its equilibrium position. For this to be the case, a continual supply of fresh reactants is required as, for example, when a reaction takes place in a flow system.

All living organisms are flow systems, consuming food and oxygen as reactants and eliminating wastes, and so it is hardly surprising that biochemical oscillators are not difficult to find: simply feel your pulse. Oscillating reactions generate the electrical signals that stimulate the heartbeat. They are also involved in the transmission of nerve signals and in the glycotic pathway in cell respiration in which glucose is broken down to release energy. Periodic variations are also common at another biological level; in the interaction of plants and animals with their neighbours and the environment. It is a basic tenet of Darwinian evolution that living organisms have the capacity to do more than replace themselves — to multiply — but predation,

food supply limitations, and disease mean that populations usually fluctuate.

A classic example is that of the snowshoe hare which is the dominant herbivore in the Canadian Arctic, feeding on the shoots of shrubs and small trees (Figure 8.4a). Cycles in the population over vast areas, from Alaska to Newfoundland, were recognized by fur trappers as early as the 1820s and are clearly seen in the detailed records of the number of skins traded by the Hudson Bay Company between 1847 and 1937 (Figure 8.4b). As with chemical oscillations, *feedback* plays a critical role, that is there are loops present by which individual steps can be turned on and off. So during periods of peak hare-abundance, plants respond to heavy grazing by producing shoots with high levels of toxins which persist for two to three years and which hares find unpalatable. The resulting food shortage leads to lower birth rates, to poor bodily condition, and consequently to a greater vulnerability to predators. As a consequence, the hare population declines, the plants become less toxic and the animals start to flourish again.

(a)

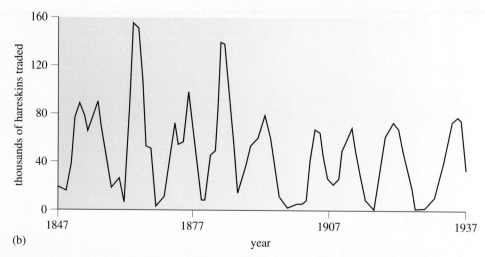

(b)

Figure 8.4   (a) The snowshoe hare (*Lepus americanus*) and (b) fluctuations in the number of hareskins traded, from 1847 to 1937.

8.4 Summary of Section 8

1 One important method for determining whether a reaction is composite involves the detection of reaction intermediates. Another involves the examination of the form of the experimental rate equation; in particular, to establish whether it is of a form different from that obtained by assuming that the reaction is elementary.

2 No matter what evidence is assembled, there is no absolute guarantee that a reaction must be elementary.

3 The rate-limiting step in a reaction mechanism is the slowest step and it controls the rate of the overall reaction.

4 If a particular elementary step in a reaction mechanism can be taken as rate-limiting then this information can be used to simplify the analysis of the mechanism. A predicted rate equation can be derived that can be compared with the experimental rate equation.

5 For a reaction with time-independent stoichiometry and a proposed reaction mechanism in which the first step is rate-limiting then the overall rate of reaction is determined by this first step and is not influenced by any of the following steps.

6 For a reaction with time-independent stoichiometry and a proposed two-step reaction mechanism in which the second step is rate-limiting then the overall rate of reaction is determined by the second step and the first step can be treated as a rapidly established pre-equilibrium.

7 As a description of how a reaction occurs a reaction mechanism is no more than a hypothetical model; it is always possible that new experimental evidence will require an established mechanism to be changed.

SUMMARY OF PART 1

The emphasis in Part 1 has been on an introduction to chemical kinetics and the analysis of reaction mechanism. A main theme has been to describe how experimental measurements of reaction rates, both as a function of concentration and temperature, can provide information on the mechanisms of chemical reactions. The coverage has been wide-ranging from fundamental well-established concepts to leading-edge research in femtochemistry.

In Section 2 we introduced the idea that chemical reactions can have reaction mechanisms that involve either a single elementary step or a sequence of such steps. The importance of elementary reactions was emphasized by looking at their properties in a little more detail. We then moved on in Sections 3 to 6 to consider experimental aspects of chemical kinetics. In particular, in Section 3 we developed a general definition for the rate of a chemical reaction. Then, following discussion in Section 4 of the factors that can influence this rate, we considered in Section 5 the various methods and strategies that can be used to establish the form of an experimental rate equation when the temperature is fixed. In Section 6 the Arrhenius equation was introduced as a means of describing the effect of temperature on the rate of a chemical reaction. This then led to a discussion of the significance of the magnitude of the Arrhenius activation energy for a given reaction.

In the remaining Sections of Part 1 we turned to a more theoretical view of chemical kinetics. In Section 7 we looked again at elementary reactions and examined models of bimolecular reactions in both the gas phase and solution. The developing area of femtochemistry was also introduced. Finally, in Section 8, we moved to a more extended discussion of reaction mechanism. We considered the type of evidence that is needed to establish that a reaction is composite and, importantly, how the concept of a rate-limiting step is used to simplify the analysis of a reaction mechanism. The latter provides a means of deriving a predicted rate equation that can be compared with experiment.

LEARNING OUTCOMES FOR PART 1

Now that you have completed Part 1 *Chemical Kinetics*, you should be able to do the following:

1   Recognize valid definitions of and use in a correct context the terms, concepts and principles in the following table (All questions).

List of scientific terms, concepts and principles used in Part 1 of this book.

Term	Page numbers	Term	Page numbers
accelerated stability study	73	integration method	50
activated complex	19	isolation method	55
activation control	84	kinetic reaction profile	22
Arrhenius activation energy, E_a	64	molecularity	78
Arrhenius A-factor	64	oscillating reaction	99
Arrhenius equation	64	overall order of reaction, n	36
Arrhenius parameters	64	partial order of reaction, α, β, γ, etc.	36
Arrhenius plot	67	physical method of analysis	41
chemical kinetics	9	plausible rate equation	45
collision theory	80	pre-equilibrium	96
completion	24	pseudo-order rate constant	57
composite reaction	17	pseudo-order rate equation	57
composite reaction mechanism	17	quenching	41
differential method	45	rate-limiting step	94
diffusion control	84	rate of a chemical reaction, J	30
elementary reaction	17	reaction coordinate	18
empirical chemical kinetics	11	reaction half-life, $t_{1/2}$	42
encounter	83	reaction intermediate	17
encounter pair	83	reaction mechanism	16
energy barrier to reaction	19	solvent cage	83
energy profile	18	stoichiometric number, ν_Y	30
experimental rate constant, k_R	35	stoichiometry	16
experimental rate equation	35	theoretical rate constant, k_{theory}	34
femtochemistry	84	theoretical rate equation	34
initial rate method	55	threshold energy, E_0	80
initial rate of reaction, J_0	28	time-independent stoichiometry	24
integrated rate equation	51	transition state	19

2 Use appropriate computer software that allows experimental data for a chemical reaction to be input and filed, and then manipulated, plotted and analysed, to obtain chemical kinetic information. (Questions 3.2, 5.1, 5.2, 5.3 and Exercises 1.1, 3.1, 5.1, 5.2 and 6.1)

3 Given data on how a quantity changes with time, determine the rate of change of this quantity with respect to time at a given instant in the time period covered by the data. (Exercise 1.1)

4 Understand what is meant by an empirical approach to chemical kinetics. (Questions 1.1, 4.1 and Exercise 5.1)

5 Draw a schematic energy profile for either an exothermic or endothermic elementary reaction and label the activation energy on this profile. (Question 2.1)

6 Appreciate that the determination of the stoichiometry of a chemical reaction, including whether it is time-independent or not, is an important step in any chemical kinetic investigation. (Question 3.1)

7 Determine at a given instant during the progress of a chemical reaction the rate of change of concentration of a reactant or product species with respect to time from a suitable kinetic reaction profile. (Question 3.2 and Exercise 3.1)

8 Given information about the initial tangent to a kinetic reaction profile, determine the initial rate of reaction. (Question 3.3)

9 Given that a reaction has time-independent stoichiometry, express the rate of reaction, J, in terms of the rate of change of concentration of each reactant and product. (Question 3.4 and Exercise 3.1)

10 Given tables of kinetic data, understand how to modify individual columns depending on the form of the table heading that is to be used. (Question 3.5 and Exercise 6.1)

11 Given an experimental rate equation for a chemical reaction, recognize whether the concept of order has meaning for the reaction and, if so, state both the partial order with respect to the individual reactants and the overall order. (Question 4.1)

12 For a chemical reaction involving just a single reactant, write down the form of the experimental rate equation in the case that the reaction is either first-order, or second-order, overall. (Question 4.2)

13 Given the form of an experimental rate equation which has a particular order, determine the units in which the experimental rate constant, or pseudo-order rate constant, would typically be expressed. (Questions 4.2, 5.2, 5.3 and Exercise 5.2)

14 Outline the main issues to be considered when using conventional techniques to investigate the kinetics of a chemical reaction.

15 Understand the definition of the term reaction half-life and use the method of successive half-lives as a preliminary check to determine whether a reaction is first-order overall or not. (Exercises 5.1, 5.2)

16 Propose a plausible rate equation and then, possibly in conjunction with an isolation method, use the differential method to determine whether the proposal is correct and, if so, establish the partial order with respect to each individual reactant. (Question 5.1 and Exercise 5.1)

17 Understand how to select and use an integrated rate equation to determine the experimental rate constant, or pseudo-order rate constant, for a chemical reaction. (Questions 5.2, 5.3 and Exercise 5.2)

18 Understand how integrated rate equations can be used either to determine, or confirm, whether a reaction involving a single reactant is first- or second-order overall. (Question 5.2 and Exercise 5.2)

19 Understand why the isolation method provides a means of simplifying the analysis of kinetic data in the case of a reaction involving several reactants. (Questions 5.3, 5.4)

20 Use information from the investigation of a chemical reaction by the initial rate method to determine the partial orders with respect to given reactant species. (Question 5.5)

21 Describe the effect of temperature on the rate of a chemical reaction and use appropriate methods to determine the Arrhenius activation energy and Arrhenius A-factor from experimental data. (Questions 6.1, 6.2 and Exercise 6.1)

22 Use the Arrhenius equation in problems concerned with the change in the rate of a chemical reaction with temperature. (Question 6.3)

23 Understand how the Arrhenius equation is used in accelerated stability studies for drugs in the pharmaceutical industry. (Question 6.4)

24 Discuss the meaning of the terms 'elementary reaction' and 'molecularity', and write down the form of the theoretical rate equation for a chemical reaction that is thought to be elementary. (Question 7.1)

25 Outline a model for a bimolecular solution-phase reaction, and distinguish between activation control and diffusion control for such a reaction.

26 Outline key features of femtochemistry, and indicate examples of its application.

27 Understand how different types of evidence can be used to determine whether a chemical reaction is composite. (Question 8.1)

28 Understand the concepts of 'rate-limiting step' and 'pre-equilibrium' in the context of the analysis of a reaction mechanism and, given the rate-limiting step in a reaction mechanism, derive the form of the predicted rate equation. (Questions 8.2, 8.3)

QUESTIONS: ANSWERS AND COMMENTS

QUESTION 1.1 (Learning Outcome 4)

Marcellin Berthelot and Péan de St Gilles found that the rate of reaction at any instant was (approximately) proportional to the concentrations of the two reactants at that instant multiplied together. The simplest possible mathematical way of representing this information is by an equation:

$$\text{rate of reaction} \propto [CH_3COOH][C_2H_5OH]$$

where the rate of reaction and the concentrations of reactants ($[CH_3COOH]$ and $[C_2H_5OH]$) are all measured at the same instant in time. If another instant in time is used then the same proportionality will hold.

The equation can be simplified a little further by introducing a proportionality constant which, for now, we will represent as k

$$\text{rate of reaction} = k[CH_3COOH][C_2H_5OH]$$

QUESTION 2.1 (Learning Outcome 5)

A schematic energy profile for the endothermic elementary reaction in Equation 2.5 is given in Figure Q.1. The difference in potential energy between products and reactants is positive, which is consistent with the endothermic nature of the reaction.

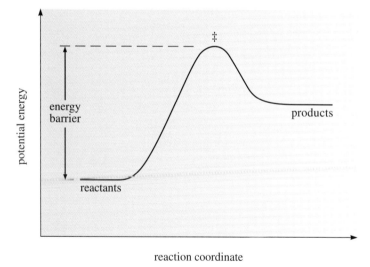

Figure Q.1   A schematic energy profile for an endothermic elementary reaction.

To be more specific in labelling, we could show on the profile that there is only a single reactant $C_6H_5CH_2Cl$, and that the products are $[C_6H_5CH_2]^+$ and Cl^-. The label ‡ depicts the transition state at which the (transient) activated complex is present.

As the schematic profile indicates, the energy barrier to reaction must be *at least* as large in magnitude as the enthalpy change for an endothermic reaction. In other words, the enthalpy change sets a *lower limit* to the energy barrier.

Questions: answers and comments

QUESTION 3.1 *(Learning Outcome 6)*

Reaction 3.1 has a known stoichiometry, which is time-independent, and the reaction essentially goes to completion. Although concise, this statement summarizes key factors to be considered in the kinetic investigation of any chemical reaction.

(Although we did not determine it, the reaction was also stated to be composite.)

QUESTION 3.2 *(Learning Outcomes 2 and 7)*

You should have obtained the values in Table Q.1. (Values are quoted to 3 significant figures.) Note that $d[ClO^-]/dt$ and $d[Br^-]/dt$ have the same value, which is also the case for $d[BrO^-]/dt$ and $d[Cl^-]/dt$.

Table Q.1 Values for the rates of change of concentration of reactants and products for Reaction 3.1 at 25 °C

	time = 1 500 s	time = 3 000 s
$\dfrac{d[ClO^-]}{dt}$ and $\dfrac{d[Br^-]}{dt}$	$-4.18 \times 10^{-7}\,mol\,dm^{-3}\,s^{-1}$	$-1.40 \times 10^{-7}\,mol\,dm^{-3}\,s^{-1}$
$\dfrac{d[BrO^-]}{dt}$ and $\dfrac{d[Cl^-]}{dt}$	$+4.18 \times 10^{-7}\,mol\,dm^{-3}\,s^{-1}$	$+1.40 \times 10^{-7}\,mol\,dm^{-3}\,s^{-1}$

As already stated in the answer to Exercise 1.1, the computer software does not provide units, and so you need to include these for yourself. It should also be remembered that although the values of the slopes of the tangents are computed by the software they are still subject to uncertainties arising from inaccuracies in the original concentration measurements.

QUESTION 3.3 *(Learning Outcome 8)*

Since Cl^- is a product, it has zero initial concentration ($t = 0\,s$, $[Cl^-] = 0\,mol\,dm^{-3}$). The question gives the coordinates of a point on the initial tangent in Figure 3.4 and so it follows that

$$\text{slope of initial tangent} = \frac{1.60 \times 10^{-3}\,mol\,dm^{-3} - 0\,mol\,dm^{-3}}{500\,s - 0\,s}$$

$$= \frac{1.60 \times 10^{-3}\,mol\,dm^{-3}}{500\,s}$$

$$= 3.20 \times 10^{-6}\,mol\,dm^{-3}\,s^{-1}$$

This quantity is equal to the *initial* value of $d[Cl^-]/dt$ and, as expected, is positive. The initial rate of reaction, J_0, is therefore equal to $+3.20 \times 10^{-6}\,mol\,dm^{-3}\,s^{-1}$.

QUESTION 3.4 *(Learning Outcome 9)*

(a) For the reaction

$$2H_2(g) + 2NO(g) = 2H_2O(g) + N_2(g)$$

$v_{H_2} = -2$, $v_{NO} = -2$, $v_{H_2O} = +2$ and $v_{N_2} = +1$.

So, using Equation 3.8 or 3.9,

$$J = -\frac{1}{2}\frac{d[H_2]}{dt} = -\frac{1}{2}\frac{d[NO]}{dt} = +\frac{1}{2}\frac{d[H_2O]}{dt} = +\frac{d[N_2]}{dt}$$

(b) For the reaction

$$S_2O_8^{2-}(aq) + 3I^-(aq) = 2SO_4^{2-}(aq) + I_3^-(aq)$$

$v_{S_2O_8^{2-}} = -1$, $v_{I^-} = -3$, $v_{SO_4^{2-}} = +2$ and $v_{I_3^-} = +1$.

Once again, using either Equation 3.8 or 3.9,

$$J = -\frac{d[S_2O_8^{2-}]}{dt} = -\frac{1}{3}\frac{d[I^-]}{dt} = +\frac{1}{2}\frac{d[SO_4^{2-}]}{dt} = +\frac{d[I_3^-]}{dt}$$

QUESTION 3.5 *(Learning Outcome 10)*

In a table of data, the column heading can be equated with any of the numbers in the column it represents. So, if we consider the first entry we have

$$\frac{[ClO^-]}{mol\,dm^{-3}} = 3.230 \times 10^{-3}$$

If *both* sides of this equation are multiplied by $mol\,dm^{-3}$, then we find

$$[ClO^-] = 3.230 \times 10^{-3}\,mol\,dm^{-3}$$

which is the expression for the concentration of ClO^-.

If instead, we take the first equation above and multiply both sides by 10^3 then

$$\frac{10^3 \times [ClO^-]}{mol\,dm^{-3}} = 10^3 \times 3.230 \times 10^{-3}$$

However, $10^3 \times 10^{-3} = 10^{3-3} = 10^0 = 1$ and it then follows that

$$\frac{10^3 \times [ClO^-]}{mol\,dm^{-3}} = 3.230$$

Thus, an appropriate table heading would be $10^3[ClO^-]/mol\,dm^{-3}$ where, for convenience, the multiplication sign has been omitted. Alternatively, it would be equally acceptable to use $[ClO^-]/10^{-3}\,mol\,dm^{-3}$. There is no convention but it is often the case that the power of ten is placed in the numerator rather than the denominator.

QUESTION 4.1 *(Learning Outcomes 4 and 11)*

For reaction (b) which represents the decomposition of a single reactant, the hypochlorite ion (ClO^-), in aqueous solution, the experimental rate equation is

$$J = k_R[ClO^-]^2$$

Thus the partial order with respect to ClO^- is 2 and the reaction is second-order in ClO^-. Since there is only a single reactant then the overall order is equal to the partial order for this reactant, that is the reaction is second-order overall.

For reaction (c) which represents the reaction between bromate ion (BrO_3^-) and bromide ion (Br^-) in acidic aqueous solution, the experimental rate equation is

$$J = k_R[BrO_3^-][Br^-][H^+]^2$$

Thus the partial order with respect to BrO_3^- is 1 (first-order), the partial order with respect to Br^- is 1 (first-order) and that with respect to H^+ is 2 (second-order). The overall order is therefore $n = 1 + 1 + 2 = 4$, that is fourth-order overall.

QUESTION 4.2 (Learning Outcomes 12 and 13)

Since the reaction involves only a single reactant, the partial order of reaction with respect to this reactant is the same as the overall order of reaction.

(a) For a first-order reaction, the partial order with respect to A will be 1, so that

$$J = k_R[A]$$

The units of k_R can be calculated from those given by $J/[A]$. Taking the units of J to be $mol\,dm^{-3}\,s^{-1}$, then the units of k_R will be $(mol\,dm^{-3}\,s^{-1})/(mol\,dm^{-3})$ which simplifies to s^{-1}.

(b) For a second-order reaction, the partial order with respect to A will be 2, so that

$$J = k_R[A]^2$$

The units of k_R can be calculated from those given by $J/[A]^2$. So, the units are $(mol\,dm^{-3}\,s^{-1})/(mol\,dm^{-3})^2$. In this expression the unit $(mol\,dm^{-3})$ can be cancelled from both the numerator and denominator to leave $(s^{-1})/(mol\,dm^{-3})$. The quantity $1/(mol\,dm^{-3})$ can be written as $mol^{-1}\,dm^3$ and so the units of k_R are $mol^{-1}\,dm^3\,s^{-1}$ or, on rearranging the order, $dm^3\,mol^{-1}\,s^{-1}$.

QUESTION 5.1 (Learning Outcomes 2 and 16)

When using computer software it is always useful, if possible, to make a preliminary estimate of a value that is to be computed.

The coordinates, with values estimated to just one decimal place, for two points on the line in Figure 5.4a are

$$\ln(J/mol\,dm^{-3}\,s^{-1}) = -11.8, \ln([NO_2]/mol\,dm^{-3}) = -5.6$$
$$\ln(J/mol\,dm^{-3}\,s^{-1}) = -14.6, \ln([NO_2]/mol\,dm^{-3}) = -7.0$$

Thus

$$\begin{aligned}
\text{slope} &= \frac{-14.6 - (-11.8)}{-7.0 - (-5.6)} \\
&= \frac{-2.8}{-1.4} \\
&= +2.0
\end{aligned}$$

To use the *Kinetics Toolkit* values of J and $[NO_2]$ from Table 5.2 are required; you may have stored these in a file already. New columns can then be created for $\ln J$ and $\ln([NO_2])$ and these quantities can be plotted against one another. (For completeness an appropriate table of data is given in Table Q.2.) You should find that the computer plot is similar to that in Figure 5.4a.

Fitting to a straight line gives

- a value of the slope equal to 2.0 within a plausible range of 1.94 to 2.06
- a value of the intercept equal to −0.68, but the statistical analysis reveals a much greater range of uncertainty with the plausible range extending from −0.31 to −1.06.

Table Q.2 Data for using the differential method to determine the partial order of reaction with respect to NO_2 for the gas-phase thermal decomposition of this compound at 300 °C

$\dfrac{[NO_2]}{\text{mol dm}^{-3}}$	$\dfrac{J}{\text{mol dm}^{-3}\,\text{s}^{-1}}$	$\ln\left(\dfrac{[NO_2]}{\text{mol dm}^{-3}}\right)$	$\ln\left(\dfrac{J}{\text{mol dm}^{-3}\cdot\text{s}^{-1}}\right)$
4.00×10^{-3}	8.32×10^{-6}	-5.521	-11.697
1.96×10^{-3}	1.90×10^{-6}	-6.235	-13.174
1.30×10^{-3}	0.88×10^{-6}	-6.645	-13.943
0.97×10^{-3}	0.48×10^{-6}	-6.938	-14.550
0.78×10^{-3}	0.30×10^{-6}	-7.156	-15.020
0.55×10^{-3}	0.16×10^{-6}	-7.506	-15.648

QUESTION 5.2 (Learning Outcomes 2, 13, 17 and 18)

Referring to Equation 5.16, the second-order integrated rate equation for the gas-phase thermal decomposition of NO_2 at 300 °C can be written in the form

$$\frac{1}{[NO_2]} = 2k_R t + \frac{1}{[NO_2]_0}$$

A plot of $1/[NO_2]$ versus time (t) should therefore be a straight line with slope $2k_R$. Such a plot is shown in Figure Q.2 and data for the plot are given in Table Q.3.

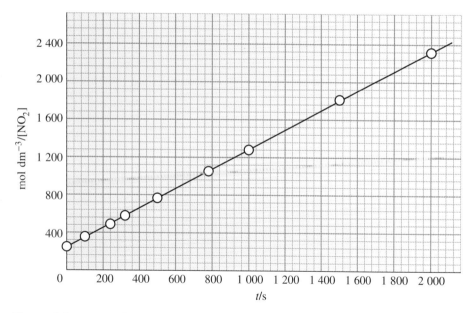

Figure Q.2 A plot of $1/[NO_2]$ versus time (t) for the gas-phase thermal decomposition of NO_2 at 300 °C.

The slope of the best straight line that can be drawn through the data points, determined using the *Kinetics Toolkit*, is

$$\text{slope} = 1.04 \, \text{dm}^3 \, \text{mol}^{-1} \, \text{s}^{-1}$$

Thus

$$2k_R = 1.04 \, \text{dm}^3 \, \text{mol}^{-1} \, \text{s}^{-1}$$

and

$$k_R = 0.52 \, \text{dm}^3 \, \text{mol}^{-1} \, \text{s}^{-1}$$

The units of k_R are those of a second-order rate constant.

The uncertainties in the value of k_R are less than $\pm 0.01 \, \text{dm}^3 \, \text{mol}^{-1} \, \text{s}^{-1}$. This is certainly better than the result that would have been obtained using the intercept in a plot of $\ln J$ versus $\ln([NO_2])$, for instance see Question 5.1.

Table Q.3
Data for determining the experimental rate constant for the gas-phase thermal decomposition of NO_2 at 300 °C

$\dfrac{\text{time}}{\text{s}}$	$\dfrac{[NO_2]}{\text{mol dm}^{-3}}$	$\dfrac{\text{mol dm}^{-3}}{[NO_2]}$
0	4.00×10^{-3}	250
100	2.83×10^{-3}	353
240	2.00×10^{-3}	500
320	1.72×10^{-3}	581
500	1.30×10^{-3}	769
780	0.94×10^{-3}	1 064
1 000	0.78×10^{-3}	1 282
1 500	0.55×10^{-3}	1 818
2 000	0.43×10^{-3}	2 326

QUESTION 5.3 *(Learning Outcomes 2, 13, 17 and 19)*

The reaction is pseudo-first-order and so a plot of $\ln([Br^-]/\text{mol dm}^{-3})$ versus time should be a straight line with a slope equal to $-k_R'$. Such a plot is shown in Figure Q.3 and data for the plot are given in Table Q.4.

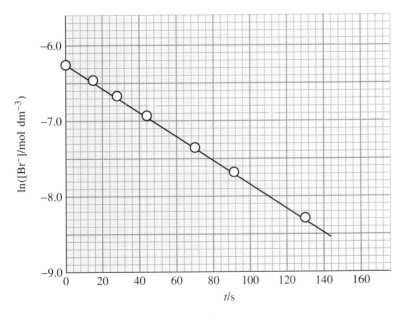

Table Q.4 Data for determining the pseudo-order rate constant k_R'

$\dfrac{\text{time}}{\text{s}}$	$\dfrac{[Br^-]}{\text{mol dm}^3}$	$\ln\left(\dfrac{[Br^-]}{\text{mol dm}^{-3}}\right)$
0	2.00×10^{-3}	-6.215
15	1.57×10^{-3}	-6.457
28	1.27×10^{-3}	-6.669
44	0.98×10^{-3}	-6.928
70	0.64×10^{-3}	-7.354
91	0.46×10^{-3}	-7.684
130	0.25×10^{-3}	-8.294

Figure Q.3 A plot of $\ln([Br^-]/\text{mol dm}^{-3})$ versus time (t). The straight line represents the best fit to all of the data points.

The slope of the straight line is

$$\text{slope} = -1.60 \times 10^{-2}\,\text{s}^{-1}$$

thus

$$-k_R' = -1.60 \times 10^{-2}\,\text{s}^{-1}$$

so that $k_R' = 1.60 \times 10^{-2}\,\text{s}^{-1}$.

The uncertainties in the value of k_R' are less than $\pm 0.02\,\text{s}^{-1}$.

QUESTION 5.4 (Learning Outcome 19)

The first step is to suggest a plausible rate equation which will be

$$J = k_R[\text{Am}]^\alpha[\text{An}]^\beta$$

It is also important to consider the relative magnitudes of the initial concentrations of the two reactants, that is $[\text{Am}]_0 = 1.12 \times 10^{-4}\,\text{mol dm}^{-3}$ (given in the question) and, from Table 5.7, $[\text{An}]_0 = 3.0 \times 10^{-6}\,\text{mol dm}^{-3}$, thus

$$\frac{[\text{Am}]_0}{[\text{An}]_0} = \frac{1.12 \times 10^{-4}\,\text{mol dm}^{-3}}{3.0 \times 10^{-6}\,\text{mol dm}^{-3}}$$

$$= 37.3$$

The amine is in large excess and the isolation method can be applied so that, to a good approximation, the concentration of the amine can be taken to be equal to its initial concentration throughout the e›]ääs   }L¿the reaction. The proposed rate equation is therefore simplified to

$$J = k_R[\text{Am}]_0^\alpha[\text{An}]^\beta$$

and this can be written as

$$J = k_R'[\text{An}]^\beta$$

where $k_R' = k_R[\text{Am}]_0^\alpha$ and is a pseudo-order rate constant.

Table 5.7 provides information for the value of J at two distinct times and, hence, two different concentrations of the anhydride. This information can be used in a trial-and-error approach to find the value of β such that $J/[\text{An}]^\beta$ is a constant. Data for cases in which $\beta = 1$ and $\beta = 2$ are summarized in Table Q.5.

Table Q.5   A trial-and-error approach to determining β

		$\beta = 1$	$\beta = 2$
$\dfrac{[\text{An}]}{\text{mol dm}^{-3}}$	$\dfrac{J}{\text{mol dm}^{-3}\,\text{s}^{-1}}$	$\dfrac{J/[\text{An}]}{\text{s}^{-1}}$	$\dfrac{J/[\text{An}]^2}{\text{mol}^{-1}\,\text{dm}^3\,\text{s}^{-1}}$
3.00×10^{-6}	11.23×10^{-9}	3.74×10^{-3}	1.25×10^3
2.33×10^{-6}	8.73×10^{-9}	3.75×10^{-3}	1.61×10^3

From the information given, albeit limited, it can be seen that $J/[\text{An}]$ is constant within experimental uncertainty and so it can be concluded that β, the partial order with respect to the anhydride (An), is 1. (It would in practice be important to check this result using data from the full kinetic reaction profile.)

QUESTION 5.5 (Learning Outcome 20)

A plausible rate equation for the *initial* rate of reaction will be

$$J_0 = k_R[\text{Am}]_0^\alpha[\text{An}]_0^\beta$$

In the two experiments in the question the initial concentration of the anhydride is fixed at $[\text{An}]_0 = 3.00 \times 10^{-6}\,\text{mol dm}^{-3}$ (but it is definitely not in excess). Thus it is the behaviour of the amine that is isolated and the proposed rate equation becomes

$$J_0 = k_R'[Am]_0^\alpha$$

where $k_R' = k_R[An]_0^\beta$ and is a pseudo-order rate constant.

Inspection of the data in Table 5.9 in the question suggests that J_0 is directly proportional to $[Am]_0$. This is confirmed by noting that the ratios $J_0/[Am]_0$ are $1.00 \times 10^{-4}\,s^{-1}$ and $1.01 \times 10^{-4}\,s^{-1}$ for *Experiments 1* and *2*, respectively. On the basis of the information provided, although limited, the partial order α, with respect to the amine (Am), is 1. (Taking into account this result and the answer to Question 5.4, the reaction between the amine and the anhydride is second-order overall, since $\alpha + \beta = 2$.)

QUESTION 6.1 *(Learning Outcome 21)*

Taking logarithms to the base 10 of the Arrhenius equation gives (Equation 6.5)

$$\log k_R = \log A - \frac{E_a}{2.303\,RT}$$

The slope of the straight line obtained by plotting $\log k_R$ versus $1/T$ is therefore

$$\text{slope} = -\frac{E_a}{2.303\,R}$$

Since we have calculated $E_a = 81.5\,kJ\,mol^{-1}$, then it follows that

$$\text{slope} = -\frac{81500\,J\,mol^{-1}}{2.303 \times 8.314\,J\,K^{-1}\,mol^{-1}}$$

where it is important to note that the value of E_a has been written in $J\,mol^{-1}$ to be consistent with the units for the gas constant, R. The value of the slope will thus be

$$\text{slope} = -4\,260\,K$$
$$= -4.26 \times 10^3\,K$$

(This value is smaller by a factor of 2.303 than the slope calculated from the Arrhenius plot using logarithms to the base e.)

QUESTION 6.2 *(Learning Outcome 21)*

If the point with a numerical value equal to 3.35 is selected on the $10^3\,K/T$ axis, then on the vertical axis, this corresponds to

$$\ln\left(\frac{k_R}{dm^3\,mol^{-1}\,s^{-1}}\right) = -6.67$$

In logarithmic (to the base e) form the Arrhenius equation is

$$\ln k_R = \ln A - \frac{E_a}{RT}$$

In more detail, this equation can be written as

$$\ln\left(\frac{k_R}{dm^3\,mol^{-1}\,s^{-1}}\right) = \ln\left(\frac{A}{dm^3\,mol^{-1}\,s^{-1}}\right) - \frac{E_a}{R}\left(\frac{1}{T}\right)$$

where $E_a = 81\,590\,J\,mol^{-1}$. Furthermore, if

$$\frac{10^3\,K}{T} = 3.35$$

then

$$\frac{1}{T} = \frac{3.35 \times 10^{-3}}{K}$$

It is convenient to deal with the last term in the detailed equation first

$$\frac{E_a}{R}\left(\frac{1}{T}\right) = \frac{81590 \text{ J mol}^{-1}}{8.314 \text{ J K}^{-1} \text{ mol}^{-1}}\left(\frac{3.35\times10^{-3}}{K}\right)$$

$$= 32.88$$

Returning to the full equation

$$-6.67 = \ln\left(\frac{A}{\text{dm}^3 \text{ mol}^{-1} \text{ s}^{-1}}\right) - 32.88$$

so that

$$\ln\left(\frac{A}{\text{dm}^3 \text{ mol}^{-1} \text{ s}^{-1}}\right) = -6.67 + 32.88$$

$$= 26.21$$

Hence, finally

$$A = 2.42 \times 10^{11} \text{ dm}^3 \text{ mol}^{-1} \text{ s}^{-1}$$

(This value is the same as that calculated using the *Kinetics Toolkit*.)

QUESTION 6.3 *(Learning Outcome 22)*

Since the initial concentration of cyclobutene is kept the same at both temperatures then the factor by which the rate of reaction increases will be the same as that by which the rate constant increases. So, using Equation 6.12

$$\ln\left(\frac{k_R(430 \text{ K})}{k_R(420 \text{ K})}\right) = \frac{137\times10^3 \text{ J mol}^{-1}}{8.314 \text{ J K}^{-1} \text{ mol}^{-1}}\left(\frac{1}{420 \text{ K}} - \frac{1}{430 \text{ K}}\right)$$

where the activation energy has been expressed in J mol^{-1}. It follows that

$$\ln\left(\frac{k_R(430 \text{ K})}{k_R(420 \text{ K})}\right) = \frac{137 \times 10^3 \text{ J mol}^{-1}}{8.314 \text{ J K}^{-1} \text{ mol}^{-1}} \times (5.54\times10^{-5} \text{ K}^{-1})$$

$$= 0.913$$

Taking the antilogarithm or calculating exp(0.913) gives

$$\frac{k_R(430 \text{ K})}{k_R(420 \text{ K})} = 2.49$$

Thus the rate of reaction at 430 K is increased by a factor of 2.49 compared to that at 420 K.

QUESTION 6.4 *(Learning Outcome 23)*

The equation

$$\ln\left(\frac{k_R(343 \text{ K})}{k_R(295 \text{ K})}\right) = \frac{E_a}{R}\left(\frac{1}{295 \text{ K}} - \frac{1}{343 \text{ K}}\right)$$

with $k_R(343 \text{ K}) = 2.49 \times 10^{-7} \text{ s}^{-1}$ and $E_a = 123.3 \text{ kJ mol}^{-1}$ can be used to determine the value of the rate constant at 295 K, that is $k_R(295 \text{ K})$. Substituting these values gives

$$\ln\left(\frac{2.49\times10^{-7}\ s^{-1}}{k_R(295\ K)}\right) = \frac{123.3\times10^3\ J\ mol^{-1}}{8.314\ J\ K^{-1}\ mol^{-1}}\left(\frac{1}{295\ K} - \frac{1}{343\ K}\right)$$

$$= 1.483\times10^4\ K\times(3.390\times10^{-3}\ K^{-1} - 2.915\times10^{-3}\ K^{-1})$$

$$= 1.483\times10^4\ K\times(4.75\times10^{-4}\ K^{-1})$$

$$= 7.04$$

So

$$\ln\left(\frac{2.49\times10^{-7}\ s^{-1}}{k_R(295\ K)}\right) = 7.04$$

and on taking the antilogarithm of both sides

$$\frac{2.49\times10^{-7}\ s^{-1}}{k_R(295\ K)} = 1.14\times10^3$$

Finally, we have

$$k_R(295\ K) = \frac{2.49\times10^{-7}\ s^{-1}}{1.14\times10^3}$$

$$= 2.18\times10^{-10}\ s^{-1}$$

If the time taken for 1% decomposition is close to $0.01/k_R$ and independent of the initial concentration of the antibiotic, then at 295 K

$$\text{time (1\% decomposition)} = \frac{0.01}{2.18\times10^{-10}\ s^{-1}}$$

$$= 4.59\times10^7\ s$$

Converting from seconds to years (that is divide by $60\times60\times24\times365$ which corresponds, in order, to a conversion to minutes, to hours, to days and finally to years) gives a value close to 1.5 years.

As long as the degradation products are not harmful, then less than 1% decomposition in a shelf-life of, say, 1 year may well be economically acceptable.

QUESTION 7.1 (Learning Outcome 24)

If they are elementary, then reactions (a), (b) and (c) are unimolecular, bimolecular and unimolecular, respectively. It thus follows that the experimental rate equations are

  (a) $J = k_R[I_2]$
  (b) $J = k_R[H_2][I]$
  (c) $J = k_R[CH_3NC]$

QUESTION 8.1 (Learning Outcome 27)

(a) Although this reaction has a relatively simple experimental rate equation, the stoichiometry is complex and so on this basis alone the reaction could not be elementary. It therefore must be composite.

(b) This reaction has been used as an example in earlier discussion. The form of the experimental rate equation is consistent with the stoichiometry and so the reaction could be elementary; it would then be a bimolecular reaction. Equally, the reaction could be composite. Without further evidence, it is not possible to decide whether this particular reaction is elementary or composite.

(c) The partial order with respect to ethanal in the experimental rate equation is fractional and equal to 3/2. The reaction must be composite.

QUESTION 8.2 *(Learning Outcome 28)*

The proposed mechanism is

$$ClO^- + ClO^- \xrightarrow{k_1} ClO_2^- + Cl^-$$

$$ClO_2^- + ClO^- \xrightarrow{k_2} ClO_3^- + Cl^-$$

If the second step had been suggested to be rate-limiting then the first step would have to be considered as a pre-equilibrium

$$ClO^- + ClO^- \underset{k_{-1}}{\overset{k_1}{\rightleftharpoons}} ClO_2^- + Cl^-$$

$$ClO_2^- + ClO^- \xrightarrow{k_2} ClO_3^- + Cl^-$$

The overall rate of reaction will be equal to that of the second step

$$J = J_2 = k_2[ClO_2^-][ClO^-]$$

For the first step, the rates of the forward and reverse reactions can be equated

$$k_1[ClO^-][ClO^-] = k_{-1}[ClO_2^-][Cl^-]$$

In other words

$$k_1[ClO^-]^2 = k_{-1}[ClO_2^-][Cl^-]$$

and the concentration of the reaction intermediate ClO_2^- is therefore given by

$$[ClO_2^-] = \frac{k_1[ClO^-]^2}{k_{-1}[Cl^-]}$$

Alternatively the equilibrium constant K_1 can be expressed as

$$K_1 = \frac{[ClO_2^-][Cl^-]}{[ClO^-][ClO^-]}$$

$$= \frac{[ClO_2^-][Cl^-]}{[ClO^-]^2}$$

and so

$$[ClO_2^-] = K_1 \frac{[ClO^-]^2}{[Cl^-]}$$

which means that $K_1 = k_1/k_{-1}$.

On substituting into the expression for the overall rate of reaction

$$J = k_2 \left(\frac{k_1}{k_{-1}} \frac{[ClO^-]^2}{[Cl^-]} \right) [ClO^-]$$

so that

$$J = \frac{k_2 k_1}{k_{-1}} \frac{[ClO^-]^3}{[Cl^-]}$$

(An alternative expression would be

$$J = k_2 K_1 \frac{[ClO^-]^3}{[Cl^-]}$$

which is equally valid.)

This predicted rate equation is in marked disagreement with experiment, as we might expect, since there is strong evidence that the second step in the proposed reaction mechanism is fast. It therefore cannot be rate-limiting.

QUESTION 8.3 *(Learning Outcome 28)*

If we assume that the first step in the proposed reaction mechanism is rate-limiting then the overall rate of reaction will be

$$J = J_1 = k_1[I_2]$$

which would clearly be in disagreement with experiment.

If we assume that the second step is rate-limiting then the first step will be a rapidly established pre-equilibrium:

$$I_2 \underset{k_{-1}}{\overset{k_1}{\rightleftharpoons}} I + I$$

$$2I + H_2 \xrightarrow{k_2} 2HI$$

The overall rate of reaction will now be equal to that of the second step

$$J = J_2 = k_2[I]^2[H_2]$$

For the pre-equilibrium

$$k_1[I_2] = k_{-1}[I][I]$$

or

$$k_1[I_2] = k_{-1}[I]^2$$

from which it is possible to find an expression for $[I]^2$, that is

$$[I]^2 = \frac{k_1}{k_{-1}}[I_2]$$

(Note that $K_1 = k_1/k_{-1}$.)

So, on substituting into the expression for J

$$J = k_2 \frac{k_1}{k_{-1}}[I_2][H_2]$$

or

$$J = k_2 K_1[I_2][H_2]$$

Either of these expressions is equivalent to the experimental rate equation where

$$k_R = k_2 \frac{k_1}{k_{-1}} = k_2 K_1.$$

To summarize, it is the *second* step in the two-step mechanism that must be rate-limiting in order to correctly predict the form of the experimental rate equation.

EXERCISES: ANSWERS AND COMMENTS

EXERCISE 1.1 (LEARNING OUTCOMES 2 AND 3)

(a) For car A, the plot of distance versus time is clearly a *straight line*; therefore, the car is travelling at a *constant speed*. If we represent distance by s and time by t, then

$$\text{speed} = \text{rate of change of distance} = \frac{\Delta s}{\Delta t}$$

and $\Delta s / \Delta t$ can be found from the slope of the straight line. This slope can be found by considering *any* convenient pair of points *on the straight line*. The best practice is to select two points that are as widely separated as possible. The uncertainties in reading the coordinates will then be much smaller fractions of the differences between the coordinates than if the two points were close together. It is also important to remember that it is the slope of the straight line *that has been drawn* that is required and so coordinates should not be taken directly from Table 1.1. (Although, in this case, they do correspond very closely with the straight line.) Using the coordinates ($t = 5$ s, $s = 90$ m) and ($t = 55$ s, $s = 980$ m) read directly from the plot in Figure 1.5a, it follows that

$$\frac{\Delta s}{\Delta t} = \frac{980 \text{ m} - 90 \text{ m}}{55 \text{ s} - 5 \text{ s}}$$

$$= \frac{890 \text{ m}}{50 \text{ s}}$$

$$= 17.8 \text{ m s}^{-1}$$

The constant speed of car A is thus calculated to be 17.8 m s^{-1} which is close to 40 miles per hour. You may well have calculated a slightly different value since the main uncertainty in the calculation lies in determining the values of the coordinates in Figure 1.5a. Indeed, given a particular time the corresponding distance can be read to within ± 5 m. This means that for the coordinates chosen the lower and upper limits for the slope are 17.6 m s^{-1} and 18.0 m s^{-1}, respectively.

For car B, the plot of distance versus time is not a straight line; it is a *curve*. The car is thus *not* travelling at a constant speed. To determine the instantaneous speed at 40 s it is necessary to draw the tangent to the curve at 40 s and determine the slope of this tangent. This is shown in Figure E.1 and the plot indicates two points from which the slope of the tangent can be calculated

$$\text{slope of tangent} = \frac{1\,430 \text{ m} - 90 \text{ m}}{60 \text{ s} - 10 \text{ s}}$$

$$= \frac{1\,340 \text{ m}}{50 \text{ s}}$$

$$= 26.8 \text{ m s}^{-1}$$

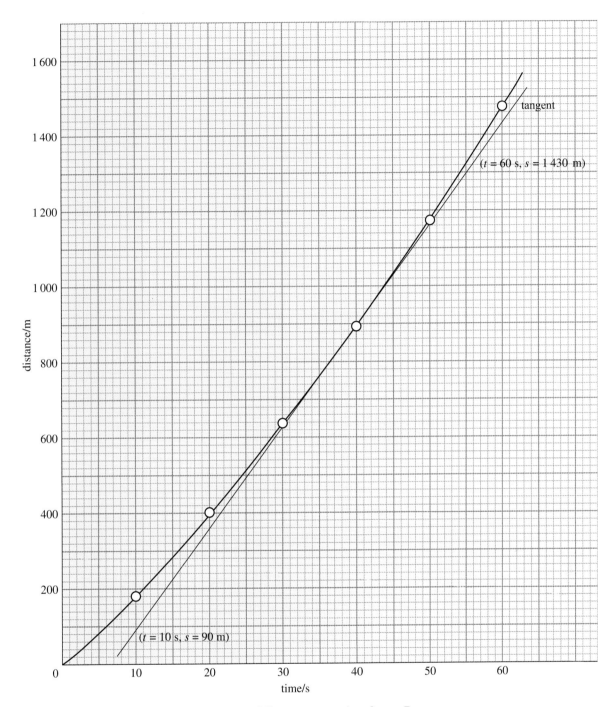

Figure E.1 The tangent at 40 s to the curve of distance versus time for car B.

As you probably found, it is not easy to draw the tangent, although the fact that the line is not too curved is helpful in this case. Your calculated value of the slope may well be different from ours; but it should at least lie within the range 24 m s⁻¹ to 30 m s⁻¹.

Car B is thus *estimated* to have an instantaneous speed at 40 s of 26.8 m s⁻¹ which is close to 60 miles per hour. The uncertainty in this estimate could well lie in the range 5% to 10%.

(b) You should have been able to enter the information given in Table 1.1 into columns in the 'Data window' of the *Kinetics Toolkit*. You should then have labelled these columns (time/s and distance/m) so that you could plot graphs similar to those in Figures 1.5a and b.

The graph of the data for car A when fitted to a *straight line* gives the following information for the slope (to 4 significant figures).

slope = 17.89 m s^{-1}

lower limit = 17.86 m s^{-1}

upper limit = 17.91 m s^{-1}

(Note that the software does *not* provide the units. You have to do this for yourself.)

The value calculated for the slope is determined from the best straight line that can be fitted to the data using statistical methods. The lower and upper limits, again determined by statistical analysis, represent a plausible range for the uncertainty in the value of the slope. This range is determined by the *experimental uncertainties* in the individual data points. (If you look at the values of the lower and upper limits using even more significant figures, you will see that they are equally disposed about the value of the slope. However, to quote the slope to more than 4 significant figures would be inappropriate given the starting data. It is reasonable to express the range for the slope as ±0.03 m s^{-1}.)

Using the *Kinetics Toolkit*, we thus find that the constant speed of car A is (17.89 ± 0.03) m s^{-1}. This result, since it is based on a statistical analysis, represents a more valid assessment of the data for car A than that carried out in part (a).

For car B you should have fitted the data to a *curve*. The slope of the tangent can then be computed at 40 s. The result (to 4 significant figures) is 26.74 m s^{-1}. This compares quite favourably with the estimate in part (a). The range of uncertainty in this value is not straightforward to calculate and is not given. However, it very much depends on the quality of the experimental data that are used in the curve fitting. For car B the data are certainly good enough for you to be confident that the slope of the tangent at 40 s is calculated to within an uncertainty of 1%. (However, at the beginning and end of the data, the uncertainty increases, but remains less than 4%.)

Calculating the slope of the tangent to a curve at a particular point is undoubtedly easier using computer software. However, this should not be taken to imply that the result is free from uncertainty.

EXERCISE 3.1 (*LEARNING OUTCOMES 2, 7 AND 9*)

Your kinetic reaction profile should be similar to that in Figure E.2.

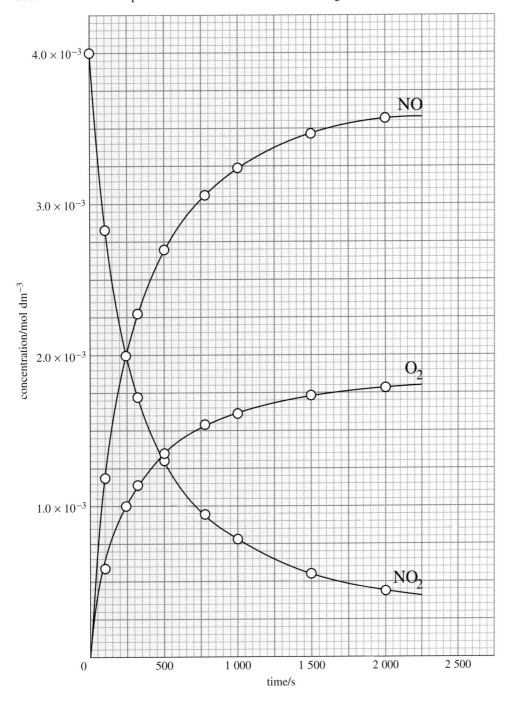

Figure E.2
A kinetic profile, obtained using data from Table 3.2, for the gas-phase decomposition of NO_2 at 300 °C. Smooth curves have been drawn through the experimental data points.

(a) By determining the slopes of the tangents to the kinetic reaction profiles of the individual species at 500 s, you should have found

$$\frac{d[NO_2]}{dt} = -1.77 \times 10^{-6} \text{ mol dm}^{-3} \text{ s}^{-1} \qquad \frac{d[NO]}{dt} = +1.76 \times 10^{-6} \text{ mol dm}^{-3} \text{ s}^{-1}$$

$$\frac{d[O_2]}{dt} = 0.88 \times 10^{-6} \text{ mol dm}^{-3} \text{ s}^{-1}$$

The rate of reaction for the decomposition is defined as (Equation 3.8)

$$J = -\frac{1}{2}\frac{d[NO_2]}{dt} = \frac{1}{2}\frac{d[NO]}{dt} = \frac{d[O_2]}{dt}$$

Hence, to a good approximation, from any of the slopes of the tangents, the rate of reaction at 500 s is given by

$$J = 0.88 \times 10^{-6}\,mol\,dm^{-3}\,s^{-1}$$

(b) Values of the rates of reaction determined at 500 s, 1 000 s and 1 500 s are summarized in Table E.1. The values can be determined, with only small variations, from any of the individual reaction profiles, but perhaps that for NO

$$J = +\frac{1}{2}\frac{d[NO]}{dt}$$

is the simplest since the slopes of the tangents are positive. (Also the kinetic reaction profile has a greater range of concentration change than that for O_2.)

Table E.1   Calculated values of the rate of reaction at 500 s, 1 000 s and 1 500 s for the gas-phase decomposition of NO_2 at 300 °C

time	J
s	$mol\,dm^{-3}\,s^{-1}$
500	0.88×10^{-6}
1 000	0.30×10^{-6}
1 500	0.16×10^{-6}

EXERCISE 5.1 (LEARNING OUTCOMES 2, 4, 15 AND 16)

(a) A plausible rate equation would be

$$J = k_R[F_2O]^\alpha$$

where α is the partial order with respect to F_2O and is also equal to the overall order of reaction.

(b) The kinetic reaction profile for the thermal decomposition of F_2O at 296.6 °C can be drawn using the *Kinetics Toolkit*; it should be similar to that in Figure E.3.

A preliminary check for first-order behaviour requires that successive half-lives are measured. From Figure E.3

$$t_{1/2}(1) \approx 200\,s$$
$$t_{1/2}(2) \approx 400\,s$$

It is not possible, given the extent of the data, to determine $t_{1/2}(3)$.

Clearly, the first two successive half-lives are not equal to one another and so the decomposition reaction *cannot* be first-order.

(c) The question specifically indicates that a *graphical* method should be used. Values of J, which are determined from

$$J = -\frac{1}{2}\frac{d[F_2O]}{dt}$$

where $d[F_2O]/dt$ is measured as the slope of the tangent to the reaction profile for F_2O at different times spanning the range of the data, are collected in Table E.2. This table also gives values of $\ln([F_2O]/mol\,dm^{-3})$ and $\ln(J/mol\,dm^{-3}\,s^{-1})$.

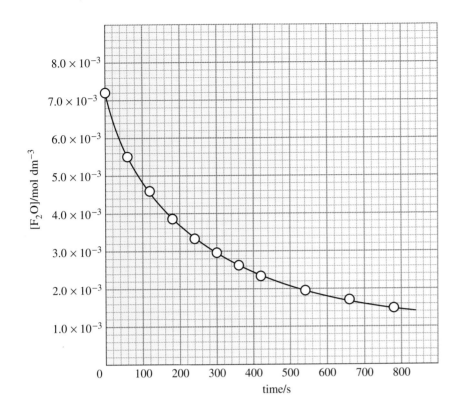

Figure E.3
A kinetic reaction profile for the thermal decomposition of F_2O at 296.6 °C. A smooth curve is drawn through the experimental data points.

Table E.2 Data required for the determination of the partial order with respect to F_2O using a graphical method

$\dfrac{time}{s}$	$\dfrac{[F_2O]}{mol\ dm^{-3}}$	$\ln\left(\dfrac{[F_2O]}{mol\ dm^{-3}}\right)$	$\dfrac{J}{mol\ dm^{-3}\ s^{-1}}$	$\ln\left(\dfrac{J}{mol\ dm^{-3}\ s^{-1}}\right)$
60	5.51×10^{-3}	-5.20	1.07×10^{-5}	-11.45
120	4.60×10^{-3}	-5.38	0.64×10^{-5}	-11.96
240	3.28×10^{-3}	-5.72	0.36×10^{-5}	-12.53
360	2.61×10^{-3}	-5.95	0.23×10^{-5}	-12.98
540	1.97×10^{-3}	-6.23	0.13×10^{-5}	-13.55

Figure E.4 shows a plot of $\ln(J/mol\ dm^{-3}\ s^{-1})$ versus $\ln([F_2O]/mol\ dm^{-3})$. This should be similar to that obtained using the *Kinetics Toolkit*. The plot is a reasonable straight line although it appears that there is more scatter in the calculated values of J in the earlier stages of reaction. The slope is 1.99 (very close to 2) with upper and lower limits of 2.24 and 1.74, respectively.

Despite the extent of the plausible range for the slope, it is still reasonable to conclude that the partial order with respect to F_2O is 2 and the decomposition reaction is second-order overall. (This conclusion would also have been reached if a second-order check had been carried out.)

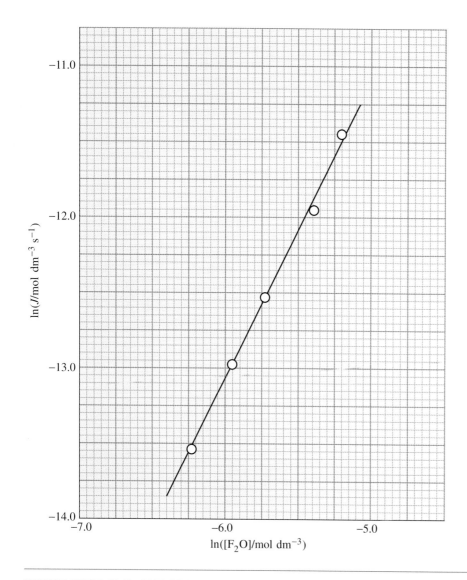

Figure E.4
A plot of $\ln(J/\text{mol dm}^{-3}\,\text{s}^{-1})$ versus $\ln([F_2O]/\text{mol dm}^{-3})$ for the gas-phase thermal decomposition of F_2O at 296.6 °C. The straight line represents the best fit to all of the data points.

The y-axis is labelled $\ln(J/\text{mol dm}^{-3}\,\text{s}^{-1})$ and the x-axis is labelled $\ln([F_2O]/\text{mol dm}^{-3})$.

EXERCISE 5.2 (LEARNING OUTCOMES 2, 13, 15, 17 AND 18)

The first step is to suggest a plausible rate equation for the decomposition reaction, that is

$$J = k_R[CH_3N_2CH_3]^\alpha$$

where α is the partial order with respect to $CH_3N_2CH_3$ and is also equal to the overall order of reaction. It is now necessary to assess whether this suggested rate equation is of the correct form and, if so, to determine the values of α and k_R. The question indicates that the 'most convenient' method of analysis should be used. This suggests (with reference to the flow diagram in Figure 5.1) starting with a preliminary check for first-order behaviour; the fact that data are provided for the decomposition reaction proceeding almost to completion means that this will be a good test. The kinetic reaction profile for $CH_3N_2CH_3$ at 332.4 °C is shown in Figure E.5.

Successive half-life steps are shown on Figure E.5. The successive half-lives are equal to one another with a value close to 460 s. This is strong evidence that the reaction is first-order, so

$$J = k_R[CH_3N_2CH_3]$$

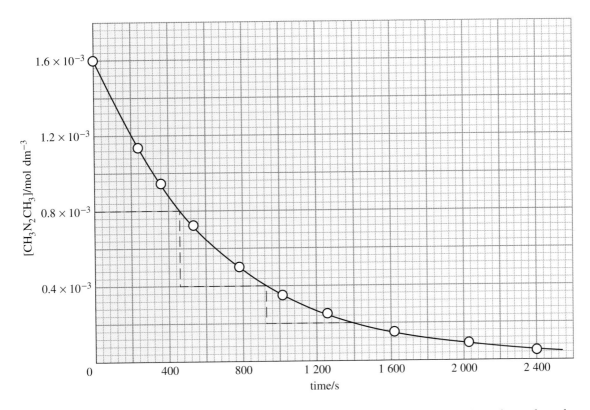

Figure E.5   The kinetic reaction profile for $CH_3N_2CH_3$ at 332.4 °C. A smooth curve has been drawn through the experimental data points.

The value of the experimental rate constant can be found using the first-order integrated rate equation

$$\ln([CH_3N_2CH_3]) = -k_R t + \ln([CH_3N_2CH_3]_0)$$

Data for plotting $\ln([CH_3N_2CH_3])$ versus time are given in Table E.3 and the plot itself is shown in Figure E.6 with the best straight line being drawn through the data points.

Table E.3   Data required for using the first-order integrated rate equation for the gas-phase decomposition of $CH_3N_2CH_3$ at 332.4 °C

$\dfrac{\text{time}}{s}$	$\dfrac{[CH_3N_2CH_3]}{\text{mol dm}^{-3}}$	$\ln\left(\dfrac{[CH_3N_2CH_3]}{\text{mol dm}^{-3}}\right)$
0	1.60×10^{-3}	-6.438
240	1.12×10^{-3}	-6.794
360	0.94×10^{-3}	-6.970
540	0.72×10^{-3}	-7.236
780	0.50×10^{-3}	-7.601
1 020	0.35×10^{-3}	-7.958
1 260	0.25×10^{-3}	-8.294
1 620	0.15×10^{-3}	-8.805
2 040	0.08×10^{-3}	-9.433
2 400	0.05×10^{-3}	-9.903

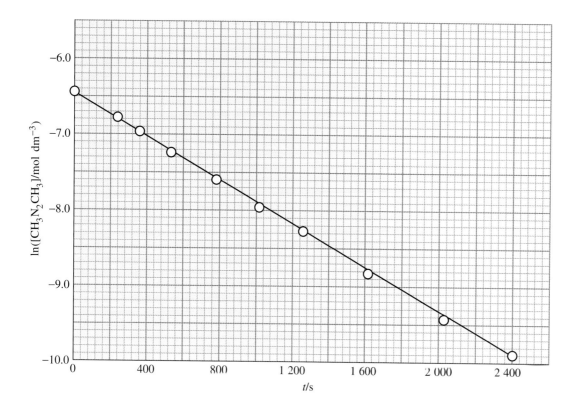

Figure E.6 A plot of $\ln([CH_3N_2CH_3]/\text{mol dm}^{-3})$ versus time (t) for the gas-phase thermal decomposition of $CH_3N_2CH_3$ at 332.4 °C.

The plot in Figure E.6, which is a very good straight line, extends to well over 50% of the complete reaction. There is thus no doubt that partial order of reaction with respect to $CH_3N_2CH_3$ is 1 and that the decomposition reaction is first-order overall.

The slope of the straight line in Figure E.6 is $-1.45 \times 10^{-3}\,\text{s}^{-1}$.

From the form of the first-order integrated rate equation, slope $= -k_R$. Thus

$$-k_R = -1.45 \times 10^{-3}\,\text{s}^{-1}$$

so that $k_R = 1.45 \times 10^{-3}\,\text{s}^{-1}$. Using the *Kinetics Toolkit*, the estimated uncertainty is $\pm 0.02 \times 10^{-3}\,\text{s}^{-1}$.

To summarize, the experimental rate equation for the gas-phase thermal decomposition of azomethane at 332.4 °C is

$$J = k_R[CH_3N_2CH_3]$$

where $k_R = 1.45 \times 10^{-3}\,\text{s}^{-1}$ at this temperature. (The overall order of this reaction was originally established in 1927.)

EXERCISE 6.1 (LEARNING OUTCOMES 2, 10 AND 21)

The Arrhenius plot, $\ln(k_R/s^{-1})$ versus reciprocal temperature, is shown in Figure E.7 for the gas-phase decomposition of chloroethane in the temperature range 670 K to 800 K. The data for this plot are summarized in Table E.4.

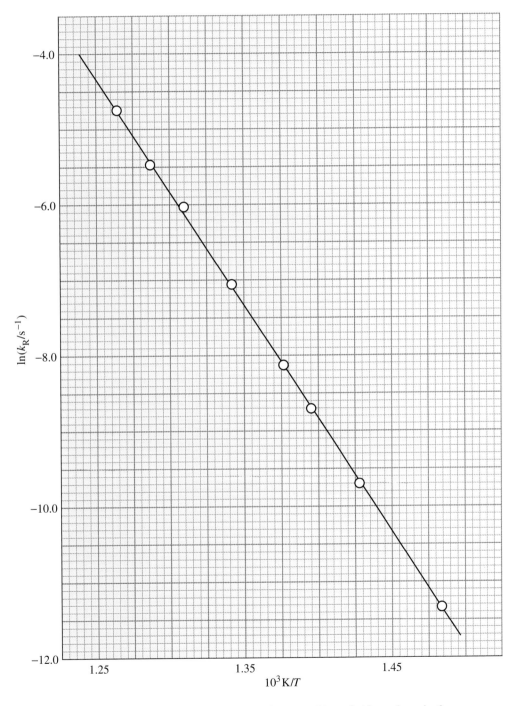

Figure E.7   An Arrhenius plot for the gas-phase decomposition of chloroethane in the temperature range 670 K to 800 K. The straight line represents the best fit to the data points.

Table E.4   Data for the Arrhenius plot in Figure E.7

$\dfrac{T}{K}$	$\dfrac{k_R}{s^{-1}}$	$\dfrac{K}{T}$	$\dfrac{10^3\,K}{T}$	$\ln(k_R/s^{-1})$
673.8	1.19×10^{-5}	1.484×10^{-3}	1.484	-11.340
699.6	6.10×10^{-5}	1.429×10^{-3}	1.429	-9.705
717.6	1.65×10^{-4}	1.394×10^{-3}	1.394	-8.710
727.2	3.00×10^{-4}	1.375×10^{-3}	1.375	-8.112
745.8	8.46×10^{-4}	1.341×10^{-3}	1.341	-7.075
764.8	2.42×10^{-3}	1.308×10^{-3}	1.308	-6.024
779.2	4.16×10^{-3}	1.283×10^{-3}	1.283	-5.482
793.2	8.41×10^{-3}	1.261×10^{-3}	1.261	-4.778

Using the *Kinetics Toolkit*, the slope and intercept of the straight line are

slope $= -2.945 \times 10^4\,K$

intercept $= 32.39$

The plausible range for the slope is from $-2.875 \times 10^4\,K$ to $-3.015 \times 10^4\,K$; for the intercept it is from 31.43 to 33.34.

Since

$$\text{slope} = -\frac{E_a}{R}$$

then it follows that

$$-\frac{E_a}{R} = -2.945 \times 10^4\,K$$

and

$$E_a = (2.945 \times 10^4\,K) \times (8.314\,J\,K^{-1}\,mol^{-1})$$
$$= 245\,kJ\,mol^{-1}$$

The plausible range for the activation energy can be calculated in a similar way; the values are $239\,kJ\,mol^{-1}$ to $251\,kJ\,mol^{-1}$. Thus from a statistical analysis of the fit of the straight line to the Arrhenius plot we can conclude that $E_a = (245 \pm 6)\,kJ\,mol^{-1}$.

Given the high temperatures at which the reaction had to be investigated in order to obtain measurable rates, it is not surprising that the activation energy has such a relatively large value.

The A-factor can be determined from the intercept which is equal to $\ln(A/s^{-1})$

$\ln(A/s^{-1}) = 32.39$

so that

$$\frac{A}{s^{-1}} = \exp(32.39) = 1.17 \times 10^{14}$$

and, finally

$A = 1.17 \times 10^{14}\,s^{-1}$

The plausible range for the A-factor can be calculated in a similar way; the values are $4.46 \times 10^{13}\,s^{-1}$ to $3.02 \times 10^{14}\,s^{-1}$. This is a relatively broad range which is a direct reflection of the uncertainty in the long extrapolation required in order to determine the intercept.

FURTHER READING

1   D. A. Johnson, *Metals and their Reactions*, The Open University and the Royal Society of Chemistry (2002), for a discussion of the subject of chemical thermodynamics.

2   P. G. Taylor, *Mechanism and Synthesis*, The Open University and the Royal Society of Chemistry (2002), for a discussion of radical chain mechanisms.

ACKNOWLEDGEMENTS

Grateful acknowledgement is made to the following sources for permission to reproduce material in this book:

Figure 1.1: Adam Butler/Associated Press; *Figure 1.2*: Associated Press; *Figure 6.2*: The Royal Swedish Academy of Sciences; *Figure 6.7a*: Z. Leszczynski/OSF; *Figure 6.8a*: Matthias Breiter/OSF; *Figure 6.8b*: Richard Packwood/OSF; *Figure 6.9*: Geoff Bryant/NHPA; *Figure 7.1*: Science and Society Picture Library; *Figure 7.6*: Courtesy of Ahmed H. Zewail; *Figure 7.7*: Courtesy of Ahmed H. Zewail; *Figure 8.2*: Oscillations and Traveling Waves in Chemical Systems by Richard J. Field and Maria Burger 1985. Reprinted by permission of John Wiley & Sons Inc. *Figure 8.4a*: Tom Ulrich/OSF.

Every effort has been made to trace all the copyright owners, but if any has been inadvertently overlooked, the publishers will be pleased to make the necessary arrangements at the first opportunity.

Part 2

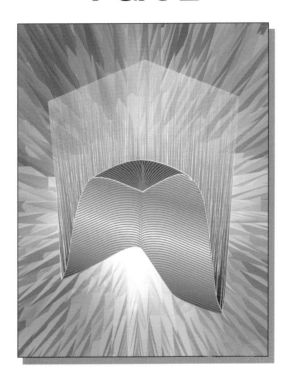

The Mechanism of Substitution

edited by Peter Taylor
from work authored by Richard Taylor

1

ORGANIC REACTIONS

1.1 Why are organic reactions important?

Organic compounds are extremely widespread in nature. But if the only organic chemicals available were the naturally occurring ones, the world would be a much poorer place (Figure 1.1). The majority of organic compounds contained in drugs, pesticides, paints, dyes, fibres and plastics are synthetic; they are not made by nature. Synthetic organic compounds are made by taking an available compound and modifying it by treatment with another chemical, in other words, by subjecting an available compound to a chemical reaction. Herein lies the strength of organic chemistry. Because one compound can be converted into another by means of a chemical reaction, the number of possible organic compounds is effectively limitless, allowing organic chemists to be truly creative.

In order to design a method for preparing a given compound, which often involves several transformations, chemists have to know a great deal about chemical reactions. They have to select suitable chemicals for effecting each desired conversion. They have to choose the best reaction conditions. Where more than one product could be formed from a reaction, they have to be able to predict with a reasonable degree of certainty which product will predominate. This may sound like an impossible task, but with further study, you too should be able to design ways of preparing fairly complicated molecules. First, though, you will have to learn about how organic reactions occur.

Figure 1.1
William Frith, *Ramsgate Sands*, 1854. A Victorian scene showing the different colours available just before the discovery of synthetic dyes.

Millions of organic reactions have been carried out successfully, and it would be an impossible task to try to remember the chemicals and reaction conditions used for each one. There are great advantages in classifying organic compounds according to structural type and functional group. Similar advantages are gained from the classification of organic reactions. By gathering together similar reactions and trying to identify the unifying features, a series of rules can be established that enable predictions to be made concerning untried reactions within that class.

Before going any further, let's define the terminology we are going to use when discussing organic reactions.

1.2 Classification of organic reactions

Chloroethane and sodium hydroxide react together to give ethanol and sodium chloride. This transformation is shown in Reaction 1.1, and as you can see, the same atoms are present on either side of the transformation arrow. In any reaction the starting materials are referred to as reactants, and they are eventually transformed into products.

$$CH_3CH_2Cl \quad + \quad Na^{+\,-}OH \quad \longrightarrow \quad CH_3CH_2OH \quad + \quad Na^+Cl \qquad (1.1)^* \quad \square$$

In this example one reactant is organic and the other is inorganic. This type of reaction is often encountered in organic chemistry, and it is conventional to differentiate between the reactants by referring to the organic reactant as the **substrate** and the inorganic reactant as the **reagent**.

● Look at the following reactions and identify the reactants and products. Are the terms substrate and reagent useful in these examples?

(i) $\quad CH_3CH_2OH \quad \xrightarrow[\text{catalyst}]{\text{acid}} \quad CH_2{=}CH_2 \quad + \quad H_2O$

(ii) $\quad CH_2{=}CH_2 \quad + \quad HBr \quad \longrightarrow \quad CH_3CH_2Br$

● In each example the reactants are shown to the left of the transformation arrow, and the products to the right. In (i) the substrate is ethanol. The transformation is effected by treating the substrate with an acidic reagent (in actual fact, sulfuric acid). The acid can be recovered at the end of the reaction and is therefore a catalyst. In (ii) there is an organic reactant and an inorganic reactant — ethene is referred to as the substrate and hydrogen bromide as the reagent.

Look back at the three reactions we have introduced in this section. They illustrate the three most common types of organic reaction. The majority of organic reactions can be classified as either substitution, elimination or addition reactions, depending on the way in which the bonding in the substrate is modified during the reaction. Let's consider each type in turn.

---

* This symbol, 🖳, indicates that the figures or structures are available in WebLab ViewerLite on the CD-ROM associated with this book.

Substitution reactions are those in which *one* atom or group in a molecule is replaced, or substituted, by another, for example

$$H-\underset{\underset{H}{|}}{\overset{\overset{H}{|}}{C}}-\underset{\underset{H}{|}}{\overset{\overset{H}{|}}{C}}-Cl \quad + \quad Na^+\,{}^-OH \quad \longrightarrow \quad H-\underset{\underset{H}{|}}{\overset{\overset{H}{|}}{C}}-\underset{\underset{H}{|}}{\overset{\overset{H}{|}}{C}}-OH \quad + \quad Na^+Cl^- \qquad (1.1)$$

The chlorine atom, which is bonded to a carbon in chloroethane, is substituted by a hydroxyl group, to produce ethanol.

Elimination reactions are those in which *two* atoms or groups are removed from a molecule, usually from adjacent carbon atoms, leading to an increased level of unsaturation, for example

$$H-\underset{\underset{H}{|}}{\overset{\overset{H}{|}}{C}}-\underset{\underset{H}{|}}{\overset{\overset{H}{|}}{C}}-OH \quad \xrightarrow[\text{catalyst}]{\text{acid}} \quad \underset{H}{\overset{H}{>}}C=C\underset{H}{\overset{H}{<}} \quad + \quad H_2O \qquad (1.2)$$

Here, a hydrogen atom and a hydroxyl group are removed from adjacent carbon atoms in ethanol, to produce ethene, which contains a double bond, and water. You will see later that alkynes, as well as alkenes, can be formed in elimination reactions.

Addition reactions are those in which *two* atoms or groups are added to a molecule containing a double or triple bond, leading to a reduced level of unsaturation, for example

$$\underset{H}{\overset{H}{>}}C=C\underset{H}{\overset{H}{<}} \quad + \quad HBr \quad \longrightarrow \quad H-\underset{\underset{H}{|}}{\overset{\overset{H}{|}}{C}}-\underset{\underset{H}{|}}{\overset{\overset{H}{|}}{C}}-Br \qquad (1.3)$$

In this example hydrogen and bromine add across the double bond in ethene (Figure 1.2) to give bromoethane, in which a single bond is left between the carbons. Addition reactions are the reverse of elimination reactions.

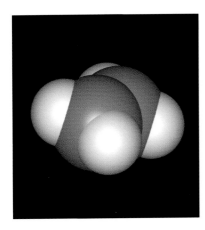

Figure 1.2
Space-filling model of ethene: notice the planar nature of this molecule.

QUESTION 1.1

Categorize the reactions (i)–(ix) as substitution, elimination or addition.

(i) CH_4 + Cl_2 $\longrightarrow$ CH_3Cl + HCl

(ii) $CH_3CH_2CH_2Br$ + K^+CN^- $\longrightarrow$ $CH_3CH_2CH_2CN$ + K^+Br^-

(iii) 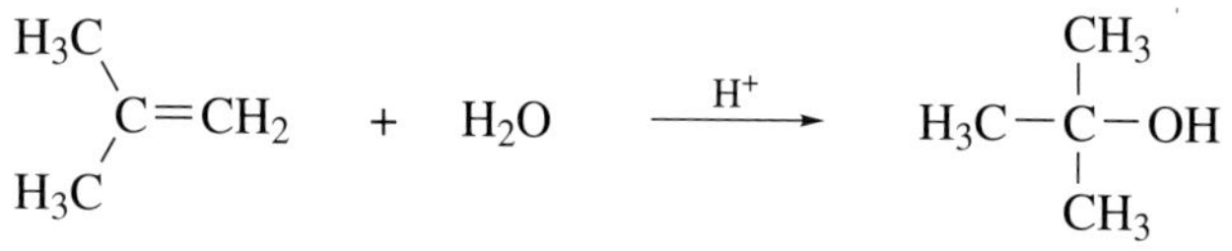 $\begin{array}{c} H_3C \\ \\ H_3C \end{array}\!\!\!C{=}CH_2$ + H_2O $\xrightarrow{\;H^+\;}$ $H_3C{-}\underset{\underset{CH_3}{|}}{\overset{\overset{CH_3}{|}}{C}}{-}OH$

(iv) $CH_3C{\equiv}CH$ + H_2 $\xrightarrow{\text{catalyst}}$ $CH_3CH{=}CH_2$

(v) $ClCH_2CH_2Cl$ + Zn $\longrightarrow$ $H_2C{=}CH_2$ + $ZnCl_2$

(vi) 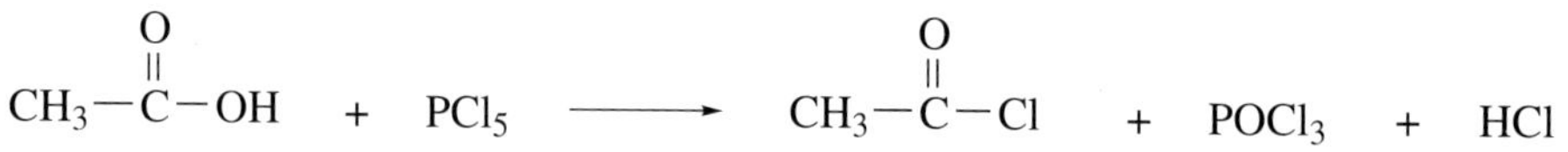 $CH_3{-}\overset{\overset{O}{\|}}{C}{-}OH$ + PCl_5 $\longrightarrow$ $CH_3{-}\overset{\overset{O}{\|}}{C}{-}Cl$ + $POCl_3$ + HCl

(vii) 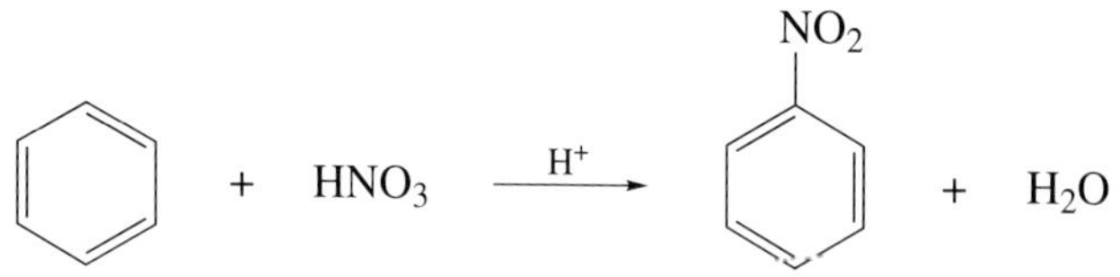   ⬡ + HNO_3 $\xrightarrow{\;H^+\;}$ ⬡NO_2 + H_2O

(viii) $CH_3{-}\overset{\overset{O}{\|}}{C}{-}H$ + HCN $\longrightarrow$ $H_3C{-}\underset{\underset{H}{|}}{\overset{\overset{OH}{|}}{C}}{-}CN$

(ix) CH_3CH_2OH + HBr $\longrightarrow$ CH_3CH_2Br + H_2O

Each of the organic reactants and products in (i) to (ix) above is available in WebLab ViewerLite on the CD-ROM associated with this book.

REACTION MECHANISMS

2.1 Reaction mechanisms: why study them?

When given the task of preparing a compound that has not been made before, organic chemists have to make a number of decisions. As well as the type of reaction, chemists must choose the substrate, reagent and reaction conditions (in particular, the solvent, temperature and time of reaction) as well as considering the use of a catalyst.

We shall look at three examples to show that these decisions are not always straightforward. Read through the examples, and after each one, try to make a note of some of the problems facing chemists when planning the reaction. Don't worry if you don't understand all the chemistry in these examples — it will be clear by the time you've finished this book!

BOX 2.1 Primary, secondary and tertiary: further nomenclature

Carbon atoms can be classified as primary, secondary or tertiary, depending upon the number of other carbon atoms attached. Thus if a carbon atom has one other carbon atom attached to it, it is known as a primary carbon (1°) and this leads to further classification, for example the primary alcohols

$$R-CH_2-X$$

primary carbon

$$CH_3-CH_2-OH, \quad CH_3-CH_2-CH_2-OH$$

primary alcohols

If a carbon atom has two other carbon atoms attached it, it is known as a secondary carbon (2°)

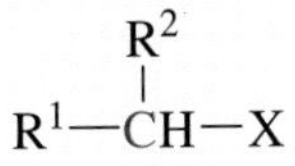

$$R^1-\overset{\overset{\displaystyle R^2}{|}}{C}H-X$$

secondary carbon

$$CH_3-\overset{\overset{\displaystyle CH_3}{|}}{C}H-OH$$

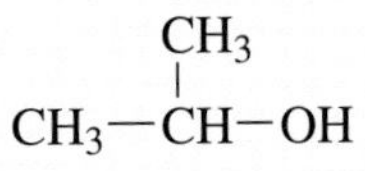

e.g. *sec*-propyl alcohol

This provides an alternative means of naming compounds. For example, the alcohol 2-propanol, a secondary alcohol, is sometimes referred to as *sec*-propyl alcohol (*sec* short for secondary).

If a carbon atom has three other carbon atoms attached it, it is known as a tertiary carbon (3°)

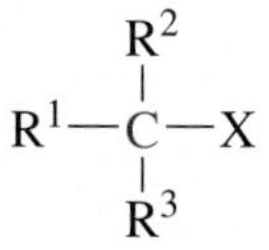

$$R^1-\underset{\underset{R^3}{|}}{\overset{\overset{R^2}{|}}{C}}-X$$

tertiary carbon

$$CH_3-\underset{\underset{CH_3}{|}}{\overset{\overset{CH_3}{|}}{C}}-OH$$

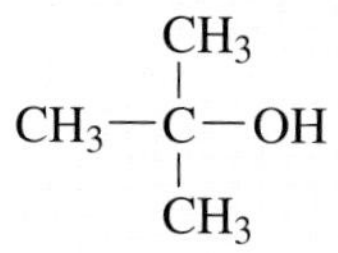

e.g. *tert*-butyl alcohol

Again this provides a means of naming compounds for example, the alcohol 2-methylpropan-2-ol is sometimes referred to as *tert*-butanol or *tert*-butyl alcohol (*tert* short for tertiary).

EXAMPLE 2.1 ●

Imagine that you have to make bromoethane for the first time, and that the only organic chemicals in the stock-cupboard are a bottle of ethanol and a cylinder of ethane. Substitution of the hydroxyl group in ethanol, or one of the hydrogen atoms in ethane, by a bromine atom will give the required product

$$CH_3CH_2OH \longrightarrow CH_3CH_2Br$$

ethanol   bromoethane

$$CH_3CH_3 \longrightarrow CH_3CH_2Br$$

ethane   bromoethane

The next question is which reagent to use.

● The reagent obviously has to contain bromine. How many inorganic reagents that contain bromine can you think of?

● You may have thought of others, but three possibilities are hydrogen bromide (HBr), bromine (Br_2), and a metal bromide such as sodium bromide (NaBr).

If you decided to investigate the reaction between ethanol and each of these three reagents in turn, you would find that only hydrogen bromide reacts with ethanol to give bromoethane

$$CH_3CH_2OH + HBr \longrightarrow CH_3CH_2Br + H_2O$$

$$CH_3CH_2OH + Br_2$$
$$CH_3CH_2OH + Na^+Br^-$$
$\Big\}$ no bromoethane formed

You might then go on to predict that ethane would also react with hydrogen bromide to give bromoethane. But this reaction would not be successful; neither hydrogen bromide nor sodium bromide reacts with ethane to give bromoethane. You would also be disappointed if you mixed ethane and bromine in the dark at room temperature, because they do not react under these conditions. However, if you persevered you might discover that ethane and bromine, when mixed in the right proportions and left in bright sunlight, do give bromoethane

$$CH_3CH_3 \quad + \quad Br_2 \quad \xrightarrow{\text{light}} \quad CH_3CH_2Br \quad + \quad HBr$$

$$CH_3CH_3 \quad + \quad HBr$$

$$CH_3CH_3 \quad + \quad Na^+Br^-$$

} no bromoethane formed, even in light

EXAMPLE 2.2 ●

In Reaction 1.1, you saw that chloroethane reacts with sodium hydroxide to give ethanol, which is a primary alcohol, and sodium chloride. Imagine that you needed to prepare not a primary alcohol, but the tertiary alcohol, 2-methylpropan-2-ol (Structure **2.1**) and that you had 2-chloro-2-methylpropane (Structure **2.2**) available. You might try to prepare the alcohol from the chloride (Structure **2.2**) using aqueous sodium hydroxide. In fact, as seen in Reaction 2.1, the product of this reaction is not the desired alcohol (Structure **2.1**) but the alkene (Structure **2.3**)

$$\begin{array}{c} CH_3 \\ | \\ H_3C-C-OH \\ | \\ CH_3 \end{array}$$

2.1 ⌨

$$\begin{array}{c} CH_3 \\ | \\ H_3C-C-Cl \\ | \\ CH_3 \end{array} \quad + \quad Na^{+\,-}OH \quad \longrightarrow \quad \begin{array}{c} H_3C \\ \diagdown \\ \diagup C=CH_2 \\ H_3C \end{array} \quad + \quad Na^+Cl^- \quad + \quad H_2O \qquad (2.1) ⌨$$

2.2   **2.3**

With aqueous sodium hydroxide, Structure **2.2** undergoes elimination rather than substitution. However, if the reaction is carried out with water alone, the desired tertiary alcohol (Structure **2.1**) is the predominant product (Reaction 2.2)

$$\begin{array}{c} CH_3 \\ | \\ H_3C-C-Cl \\ | \\ CH_3 \end{array} \quad \xrightarrow{H_2O} \quad \begin{array}{c} CH_3 \\ | \\ H_3C-C-OH \\ | \\ CH_3 \end{array} \quad + \quad \begin{array}{c} H_3C \\ \diagdown \\ \diagup C=CH_2 \\ H_3C \end{array} \qquad (2.2) ⌨$$

2.2   **2.1** (83%)   **2.3** (17%)

● ●

EXAMPLE 2.3 ●●●●●●●●●●●●●●●●●●●●●●●●●●●●●●●●

Imagine you wanted to prepare octan-2-ol with the configuration shown in Structure **2.4**.

You might decide that, by analogy with Example 2.2, conditions could be found for preparing Structure **2.4** from the corresponding chloroalkane (Structure **2.5**). However, although Compound **2.5** does undergo substitution with water or aqueous sodium hydroxide to give an optically active product, it does not lead to Compound **2.4**, but to the enantiomer (Structure **2.6**) of the desired compound (Reaction 2.3). In order to prepare the alcohol Structure **2.4** you would have to start with the enantiomer of the chloroalkane (Structure **2.7**) as shown in Reaction 2.4. A space-filling model of Structures **2.7** and **2.4** is shown in Figure 2.1.

$$C_6H_{13}$$
$$HO-\overset{\displaystyle C_6H_{13}}{\underset{\displaystyle H}{C}}-CH_3$$

2.4

$$Cl-\overset{\displaystyle C_6H_{13}}{\underset{\displaystyle H}{C}}-CH_3 \quad + \quad Na^{+-}OH \quad \longrightarrow \quad H_3C-\overset{\displaystyle C_6H_{13}}{\underset{\displaystyle H}{C}}-OH \quad + \quad Na^+Cl^- \quad\quad (2.3)$$

2.5   **2.6**

$$H_3C-\overset{\displaystyle C_6H_{13}}{\underset{\displaystyle H}{C}}-Cl \quad + \quad Na^{+-}OH \quad \longrightarrow \quad HO-\overset{\displaystyle C_6H_{13}}{\underset{\displaystyle H}{C}}-CH_3 \quad + \quad Na^+Cl^- \quad\quad (2.4)$$

2.7   **2.4**

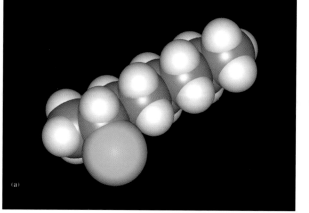
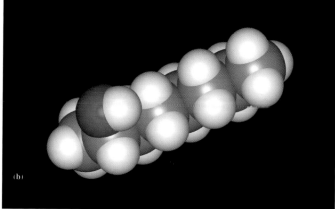

Figure 2.1 Space-filling models of (a) Structure **2.7** and (b) Structure **2.4**. Notice that the Cl hasn't been simply replaced by the OH since the stereochemistry has changed.

●●●

Now read through Examples 2.1, 2.2 and 2.3 again and jot down some of the problems that organic chemists must tackle when planning a reaction.

Your list might include some of the following:

1 Choice of substrate structure, especially when stereochemical considerations are involved (Example 2.3);

2 Choice of reagent, which is not as straightforward as it sounds (Examples 2.1 and 2.2);

3 Ensuring that the reaction conditions are adequate: some reactants can be mixed and reaction won't occur until they are heated or irradiated (Example 2.1);

4 Ensuring that the required type of reaction occurs where more than one type is possible: substitution and elimination often compete with each other (Example 2.2).

In most of the reactions that we have described here, the substrate has been modified by reaction at the functional group, and the hydrocarbon framework has remained unchanged. In Example 2.1, ethanol reacted with hydrogen bromide at the site where the hydroxyl group is attached to produce bromoethane

$$CH_3CH_2OH + HBr \longrightarrow CH_3CH_2Br + H_2O$$

Similarly the chloro group was involved in reaction in both Example 2.2 and Example 2.3. However, the second reaction within Example 2.1 showed that under certain conditions the hydrogen in a C−H bond can undergo substitution

$$CH_3CH_3 + Br_2 \xrightarrow{\text{light}} CH_3CH_2Br + HBr$$

If you had to prepare the more complicated molecule bromocyclohexylmethane (Structure **2.8**), which of the two routes shown would you expect to give the better chance of success?

Route 1

CH₃   + Br₂ $\xrightarrow{\text{light}}$ CH₂Br   + HBr

2.9   **2.8**

Route 2

CH₂OH   + HBr $\longrightarrow$ CH₂Br   + H₂O

2.10   **2.8**

Route 1 involves the substitution of a hydrogen atom on the methyl group in methylcyclohexane (Structure **2.9**) by bromine. However, there are 14 hydrogen atoms in the molecule (Figure 2.2), and other brominated products are possible, such as the four shown below. In practice, the direct bromination reaction shown in Route 1 would probably give a mixture of all of the possible brominated products. It would then be difficult to separate the products, and chemicals would be wasted.

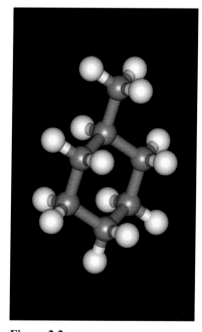

Figure 2.2
Methylcyclohexane (Structure **2.9**) contains 14 hydrogens. 🖥

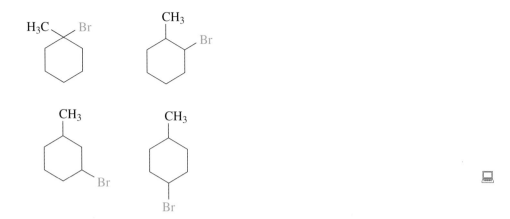

Route 2 uses cyclohexylmethanol (Structure **2.10**), which contains a hydroxyl group in the position at which bromine substitution is required. Simple primary alcohols, such as ethanol, react with HBr to give the corresponding bromoalkane, and if this reaction can be extended to other primary alcohols, such as Structure **2.10**, it would provide a good way of preparing bromocyclohexylmethane (Structure **2.8**).

It is an obvious advantage to be able to predict the site within a molecule at which a reaction will take place and this can be achieved by ensuring the reaction occurs at a specific functional group.

By now you should be able to see that in order to plan a reaction, even a fairly simple one, quite a lot of detailed knowledge might be required. In order to carry out a reaction successfully, it is necessary to understand how and why that reaction takes place. It is necessary to know about the reaction on a molecular level; to know as much as possible about the detailed sequence of elementary reactions followed by the reactant molecules as they are converted into product molecules. As you saw in Part 1, this is called the **reaction mechanism**.

You will be relieved to know that by studying the mechanisms of simple organic reactions it is possible to draw up a set of ground rules, which can be used predictively. By studying simple substitution reactions, such as the conversion of typical primary, secondary or tertiary alcohols into bromoalkanes, it is possible to predict the best reagents and conditions for any particular substitution reaction, such as the conversion of cyclohexylmethanol into bromocyclohexylmethane.

By the time you have finished this book you will be able to explain all the observations arising from Examples 2.1, 2.2 and 2.3. You will also be able to use in a predictive sense the information you have learned. But, it is when you examine organic synthesis, that the true worth of the study of reaction mechanisms becomes apparent.

So let's start our examination of organic reaction mechanisms by looking at the ways in which covalent bonds can be broken and made during a reaction.

2.2 Breaking and making covalent bonds

The bonds in organic molecules are generally of the two-electron covalent type. Organic reactions involve the breaking and making of covalent bonds.

There are two distinct ways in which a covalent bond can break; in a symmetrical or an unsymmetrical fashion.

In a *symmetrical* cleavage, when a covalent bond breaks, each of the constituent atoms retains one electron (represented by a dot or circle) from the bond that is cleaved

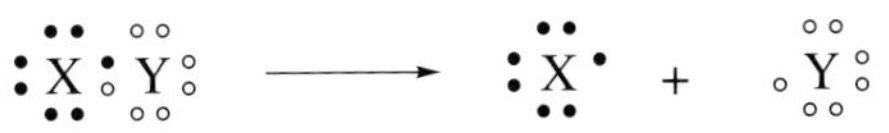

$$:\!X\!:\!Y\!: \longrightarrow :\!X^{\bullet} + {}^{\circ}\!Y\!:$$

or

$$X\!:\!Y \longrightarrow X^{\bullet} + Y^{\circ}$$

Such a process is known as **bond homolysis**, and the species produced, which each possess one unpaired electron, are called **radicals**.

In an *unsymmetrical* cleavage, when a covalent bond breaks, one of the constituent atoms gains both electrons from the bond

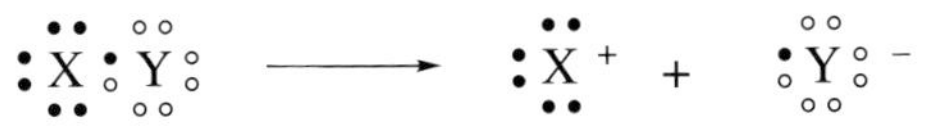

$$:\!X\!:\!Y\!: \longrightarrow :\!X^{+} + {}^{\bullet}\!Y\!:^{-}$$

or

$$X\!:\!Y \longrightarrow X^{+} + Y\!:^{-}$$

This type of process, known as **bond heterolysis**, leads to charged species: a negatively charged anion (usually written Y^-), which has both the electrons from the cleaved bond, and a positively charged cation, X^+.

○ In the last equation above, why does X carry a positive charge and Y a negative charge?

◐ The $X-Y$ bond consists of two electrons, one from X and one from Y. Since both electrons go to Y, the X cation has lost control of the electron it supplied, and so it is overall positively charged, whereas Y gains this electron, and so becomes negatively charged.

The molecule XY could equally well yield the X^- anion and the Y^+ cation, if the two electrons ended up on the X atom. Which atom forms the anion, and which forms the cation, depends on the nature of the atoms X and Y.

Reactions in which bonds are broken in a homolytic manner (that is, by the movement of single electrons) are called **radical reactions**. Reactions in which bonds are broken in a heterolytic manner (that is, by the movement of pairs of electrons) are called **ionic reactions**.

Let's look at the main features of each type of reaction.

2.2.1 Radical reactions

We have already introduced two reactions that proceed by a radical mechanism

$$CH_4 + Cl_2 \longrightarrow CH_3Cl + HCl$$

$$CH_3CH_3 + Br_2 \xrightarrow{\text{light}} CH_3CH_2Br + HBr$$

These simple equations belie the complex mechanism of such reactions. Let's consider the chlorination of methane, for example. This reaction proceeds by a multi-step process, the first step being homolysis of the chlorine molecule, which is brought about by heating or irradiating the reaction mixture

$$:\ddot{\underset{..}{Cl}}:\overset{\circ\circ}{\underset{\circ\circ}{Cl}}:\quad\xrightarrow{\text{heat or light}}\quad :\ddot{\underset{..}{Cl}}\cdot\quad+\quad\overset{\circ\circ}{\underset{\circ\circ}{\cdot Cl}}:$$

● Are the chlorine radicals charged?

● No. Each chlorine radical has seven valence electrons, which means that chlorine radicals are the same as chlorine atoms. It is normal to show only the unpaired electron on a radical (in this case, Cl•).

Chlorine radicals (like most radicals) are very reactive, because of the unpaired electron. When such a radical reacts with a molecule of methane, a carbon–hydrogen bond is broken and a methyl radical and a hydrogen chloride molecule are formed

$$Cl\bullet \quad + \quad H-\underset{\underset{H}{|}}{\overset{\overset{H}{|}}{C}}-H \quad\longrightarrow\quad Cl-H \quad + \quad \bullet\underset{\underset{H}{|}}{\overset{\overset{H}{|}}{C}}-H$$

Because of the unpaired electron on the carbon atom, the methyl radical is also very reactive, and it reacts with a chlorine molecule to form chloromethane and another chlorine radical

$$H_3C\bullet \quad + \quad Cl-Cl \quad\longrightarrow\quad CH_3-Cl \quad + \quad Cl\bullet$$

This chlorine radical can in turn react with another methane molecule, to give a methyl radical, and so the process continues. This type of self-perpetuating reaction is known as a **chain reaction**.

Other processes involving radicals occur during the halogenation of methane, and further complicate the mechanism, but these need not concern you now. The important point to note is that in radical halogenation reactions, a carbon–hydrogen bond can be replaced by a carbon–halogen bond. Radical halogenations are generally unselective; because all types of C—H bond, although not identical, are of similar reactivity. Therefore, with more complicated hydrocarbons a mixture of all possible products is usually obtained.

● How many products are possible from the monochlorination (that is, the substitution of only one H by Cl) of propane (whose structure is shown in Figure 2.3)?

● Two products are possible, 1-chloropropane and 2-chloropropane, and they are obtained in roughly equal amounts

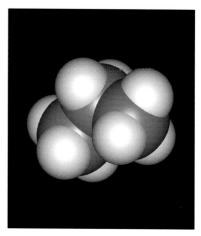

Figure 2.3
Propane contains 8 hydrogens. 🖳

$$CH_3CH_2CH_3 \quad\xrightarrow[\text{light}]{Cl_2}\quad CH_3CH_2CH_2Cl \quad + \quad CH_3\underset{\overset{|}{Cl}}{\overset{\overset{Cl}{|}}{C}}HCH_3$$

1-chloropropane (45%)   2-chloropropane (55%)

The number of possible products increases with molecular size. As we saw earlier, methylcyclohexane (Structure **2.9**) can give rise to five monobrominated products. A mixture of products is often a feature of radical mechanisms.

Because a high yield of one particular compound is usually needed, and the separation of product mixtures is often difficult, radical halogenations are rarely used with substrates other than simple hydrocarbons. This is not to say that this reaction type is unimportant: the chlorination of methane is a major industrial process. Chloromethane is not the only product obtainable: if the ratio of the reactants and the reaction conditions are varied, dichloromethane (CH_2Cl_2), trichloromethane (chloroform, $CHCl_3$) and tetrachloromethane (carbon tetrachloride, CCl_4) can all be produced.

2.2.2 Ionic reactions

The majority of the reactions that you've met so far in this book have been ionic reactions, such as Reactions 2.5–2.8.

$$CH_3-CH_2-Cl \quad + \quad Na^+ \; {}^-OH \quad \longrightarrow \quad CH_3-CH_2-OH \quad + \quad Na^+ \; Cl^- \tag{2.5}$$

$$CH_2{=}CH_2 \quad + \quad HBr \quad \longrightarrow \quad CH_3-CH_2-Br \tag{2.6}$$

$$CH_3-CH_2-CH_2-Br \quad + \quad Na^+ \; {}^-CN \quad \longrightarrow \quad CH_3-CH_2-CH_2-CN \quad + \quad Na^+ \; Br^- \tag{2.7}$$

$$H_3C-\underset{\underset{CH_3}{|}}{\overset{\overset{CH_3}{|}}{C}}-Cl \quad + \quad Na^+ \; {}^-OH \quad \longrightarrow \quad \underset{H_3C}{\overset{H_3C}{>}}C{=}CH_2 \quad + \quad Na^+ \; Cl^- \quad + \quad H_2O \tag{2.8}$$

So, perhaps we should ask ourselves what features distinguish such ionic reactions from radical reactions, such as the chlorination of methane.

1   Ionic reactions involve the movement of pairs of electrons. For example, in Reaction 2.5 both electrons in the C—Cl bond go to the chlorine, to give Cl⁻, not Cl•.

2   In the radical halogenation reaction a carbon–hydrogen bond is replaced, whereas in Reactions 2.5–2.8 the reactions occur at functional groups and the carbon–hydrogen bonds are unaffected, although in the elimination Reaction 2.8 a hydrogen atom adjacent to the functional group is also lost.

It is this second factor, that ionic reactions occur at the site of a functional group, that makes ionic reactions so useful. We shall therefore concentrate on ionic reactions for the rest of the book. The key point can be stated as follows:

> When an organic molecule reacts by an ionic mechanism, the hydrocarbon framework can be considered to be inert, and the possible sites of reaction (or reactive sites) within the molecule are located at or around the functional groups.

⬤ Identify the reactive sites in the following molecules

$$\overset{\displaystyle OH}{\underset{\displaystyle }{|}}$$

(a) CH_3CHCH_3

(b) $CH_3CH=CHCH_3$

⬤ The functional groups provide the reactive sites, and so the alcohol (C–OH) functional group in (a), and the alkene (C=C) functional group in (b), would be expected to participate in ionic reactions. Reactions 2.9 and 2.10. show this to be correct.

$$\overset{OH}{\underset{|}{CH_3CHCH_3}} \;+\; HBr \;\longrightarrow\; \overset{Br}{\underset{|}{CH_3CHCH_3}} \;+\; H_2O \qquad (2.9) \;\;▯$$

$$CH_3CH=CHCH_3 \;+\; HBr \;\longrightarrow\; \overset{Br}{\underset{|}{CH_3CH_2CHCH_3}} \qquad (2.10) \;\;▯$$

Although reactive sites can be readily identified, it is important to remember that the correct reagent has to be found for carrying out the desired conversion. In order to be successful, a proposed reaction has to be thermodynamically *and* kinetically favoured. For example, ethanol has a hydroxyl group, and sodium bromide is an ionic reagent but, as you saw earlier, these two compounds do not react to form bromoethane, whereas ethanol and hydrogen bromide do.

In the next section, we shall concentrate on the mechanisms of substitution reactions, and you will see why ionic reactions occur at functional groups, and why ethanol reacts with HBr and not with NaBr.

2.3 Summary of Sections 1 and 2

1 Organic reactions are used to prepare many useful compounds from readily available starting materials.

2 The three major classes of organic reactions are substitution reactions, elimination reactions and addition reactions.

3 In order to plan successful organic reactions, a detailed knowledge of organic mechanisms is required.

4 Some reactions, such as the halogenation of aliphatic hydrocarbons, proceed by radical mechanisms. Radicals possess an unpaired electron resulting from bond homolysis. They are generally uncharged.

5 In general, the radical halogenation of alkanes is indiscriminate, with substitution occurring at every carbon–hydrogen bond. This limits the usefulness of this type of reaction to small hydrocarbon molecules. However, some such reactions (for example, the chlorination of methane) are of great industrial importance.

6 Other reactions proceed by ionic mechanisms. In ionic reactions, bonds are broken heterolytically. The hydrocarbon framework is unreactive, and the functional groups provide the possible sites of reaction. Whether or not reaction occurs depends on the choice of substrate and reagent.

QUESTION 2.1

Classify the following reactions as substitution, addition or elimination. Decide whether ionic or radical mechanisms are operating, and give reasons for your decisions.

(i) CH_3Br + Na^+ ^-OH $\longrightarrow$ CH_3OH + Na^+ Br^-

(ii)

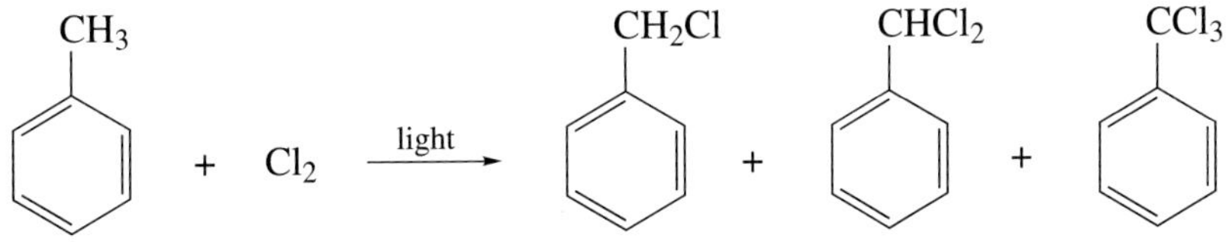

(iii)

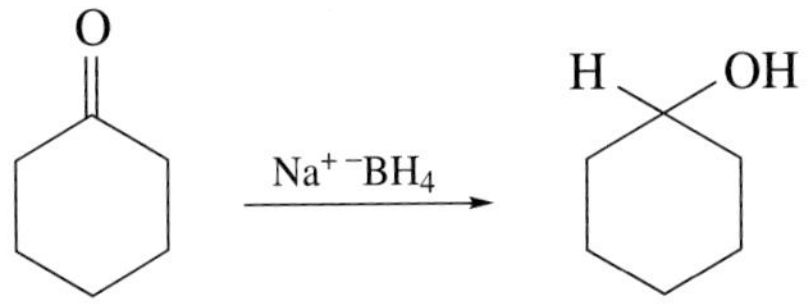

(iv) $CH_3C\equiv CCH_3$ + HBr $\longrightarrow$ $CH_3CH= \overset{Br}{\underset{|}{C}}CH_3$

(v)

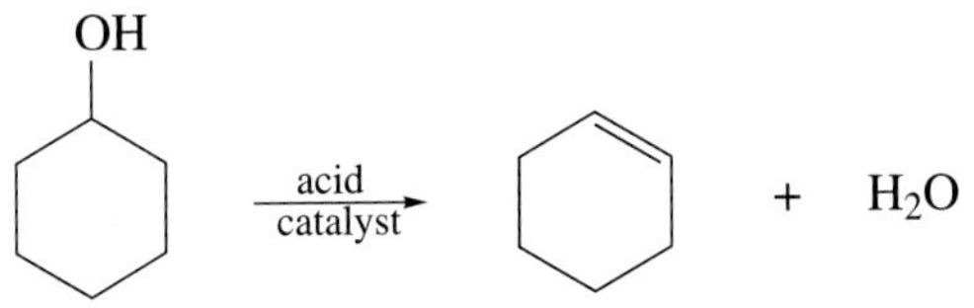

Each of the organic reactants and products in (i) to (v) above is available in WebLab ViewerLite on the CD-ROM associated with this book.

IONIC SUBSTITUTION REACTIONS

3

3.1 Nucleophiles, electrophiles and leaving groups

In Sections 1 and 2 we discussed a few ionic substitution reactions, and several more are listed in Table 3.1. Examples (a)–(d) are typical of the substitution reactions that we shall be discussing in this part of the book.

Table 3.1 Some typical ionic substitution reactions. Reaction (e) is shown happening in the laboratory in Figure 3.1. All organic species in (a)–(f) are available in WebLab

(a) $\qquad CH_3Br \ + \ Na^{+\ -}OH \ \longrightarrow \ CH_3OH \ + \ Na^+Br^-$

(b) $\quad CH_3CH_2CH_2Br \ + \ K^{+\ -}CN \ \longrightarrow \ CH_3CH_2CH_2CN \ + \ K^+Br^-$

(c) $\quad CH_3CH_2Cl \ + \ Na^{+\ -}OCH_2CH_3 \ \longrightarrow \ CH_3CH_2OCH_2CH_3 \ + \ Na^+Cl^-$

(d) $\qquad CH_3CH_2OH \ + \ HBr \ \longrightarrow \ CH_3CH_2Br \ + \ H_2O$

(e) [benzene ring] $\ + \ HNO_3 \ \xrightarrow{\ H^+\ }$ [benzene ring with NO$_2$] $\ + \ H_2O$

(f) $\qquad CH_3COOH \ + \ PCl_5 \ \longrightarrow \ CH_3COCl \ + \ POCl_3 \ + \ HCl$

● Look at Reactions (a)–(d) in Table 3.1. What features are common to all four reactions?

● All the substrates contain a functional group (Br, Cl or OH) joined to an aliphatic carbon atom by a covalent bond. In each reaction the functional group is replaced by a group (OH, CN, OCH$_2$CH$_3$ or Br) supplied by the reagent.

This behaviour is typical of a reaction proceeding by an ionic mechanism. Let's look at Reaction (a) in greater detail, in order to see the bonding changes that are occurring.

Another way of depicting this reaction is to show all the outer-shell electrons, as in the following Lewis structures (oxygen is in Group VI and bromine in Group VII of the Periodic Table)

Figure 3.1
The reaction of benzene with nitric acid: brown fumes of NO$_2$ are formed as a by-product.

Remember that each covalent bond (for example, C—H) comprises two shared electrons, so this reaction can be drawn as follows

During the course of the reaction the oxygen atom of the hydroxide anion $^-$OH becomes covalently bonded to the carbon atom, and the bond from carbon to bromine breaks, giving a bromide anion Br^-. The sodium ion is merely a spectator as far as the bond-making and bond-breaking are concerned, which is why it is often referred to as a **spectator ion**.

For the bromide anion to be formed when the carbon–bromine bond breaks, both electrons from the bond must be donated to bromine. Similarly, in order to form a covalent bond to carbon, the oxygen atom must donate two electrons.

Bearing in mind that this reaction proceeds by an ionic mechanism (that is, the bond-making and bond-breaking involve the movement of electron pairs), you should now be able to see the electron movements needed in order to transform the reactants into the products. One of the non-bonded electron pairs on the hydroxyl oxygen atom (shown in red) can form the new bond to the carbon atom, and in order to generate the bromide anion, the C—Br bond breaks heterolytically, with both electrons (shown in green) going to the bromine atom.

We can get an indication of why this reaction occurs if we look at the polarity of the bonds in the substrate. Carbon and hydrogen have similar electronegativities so there is little charge build-up in the C—H bonds. However, there is a larger electronegativity difference between the carbon and the bromine so the bond is polarized in the sense $C^{\delta+}$—$Br^{\delta-}$. Thus the negatively charged hydroxide ion is attracted to the positively (albeit slightly, see Figure 3.2 overleaf) charged carbon bearing the bromine:

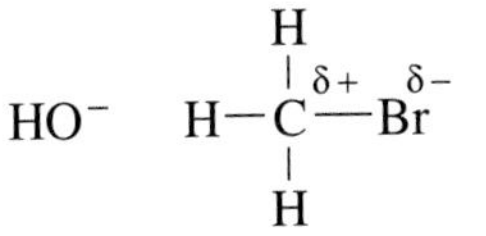

Here, then, is one rationalization of the observation that reactive sites in molecules are normally associated with functional groups. The bromine atom confers polarity on an otherwise effectively non-polar hydrocarbon framework, and ionic reagents might be expected to react at polar sites for electrostatic reasons.

The attacking group, $^-$OH in this example, that forms a new bond to the carbon is known as the **nucleophile**. Nucleophile means 'lover of nuclei', and nucleophiles try to form a covalent bond to another atom. To do this a nucleophile must possess at least one non-bonded pair of electrons, which ultimately forms a new covalent bond. As we have seen, the nucleophile is attracted to the positive charge on the

Figure 3.2
The charge density distribution
in (a) the hydroxide ion and
(b) bromomethane. The red areas of
the surface indicate regions of high
electron density, and the blue areas
indicate low electron density.

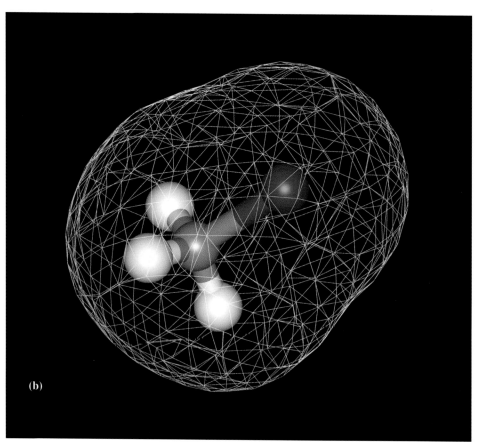

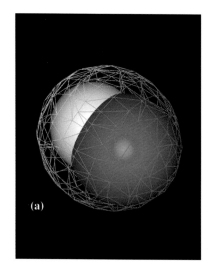

(a)　　　　　(b)

electrophilic atom. The term **electrophile** means 'lover of electrons' and most
electrophilic atoms are positively charged or positively polarized, just as most
nucleophiles are negatively charged or negatively polarized. The group that is lost,
bromine in this example, is known as the **leaving group**.

> To help you keep track of what is what, in the next few sections we shall
> draw nucleophiles in red, electrophilic centres in blue and leaving groups
> in green.

Which is the nucleophile, and which is the electrophilic atom, in the following
transformation? Which is the leaving group?

$$^-CN \ + \ H-\underset{\underset{H}{|}}{\overset{\overset{H}{|}}{C}}-\underset{\underset{H}{|}}{\overset{\overset{H}{|}}{C}}-Br \ \longrightarrow \ H-\underset{\underset{H}{|}}{\overset{\overset{H}{|}}{C}}-\underset{\underset{H}{|}}{\overset{\overset{H}{|}}{C}}-CN \ + \ Br^-$$

Cyanide ($^-$CN) is the nucleophile, and the carbon atom attached to bromine is
the electrophilic atom. The bromine (Br) is the leaving group.

As expected in an ionic process, the position of the functional group determines the
site of reaction. In this case the bromine atom renders the adjacent carbon atom
electrophilic, since the carbon–bromine bond is polarized in the sense

$$\overset{\delta+ \quad \delta-}{\underset{/}{\overset{\backslash}{C}}-Br}$$

polarization of
carbon–bromine bond

We can generalize this type of substitution for any nucleophile and substrate. To simplify things we have ignored the cation for the time being, as it doesn't participate in the reaction. The general equation for this type of substitution at a saturated carbon atom can be written as follows, ignoring the spectator ions

$$Nu^- + \quad -\overset{|}{\underset{|}{C}}-X \quad \longrightarrow \quad -\overset{|}{\underset{|}{C}}-Nu \quad + \quad X^- \qquad (3.1)$$

In this general form of the reaction, the nucleophile * Nu^- donates a pair of electrons to the electrophilic carbon to form a new bond and the leaving group, X, takes both electrons in the C—X bond with it when it leaves, thus forming a negatively charged anion, X^-.

● Predict the substitution product when the nucleophile Br^- reacts with 2-iodopropane (Structure **3.1**) which contains the iodine leaving group.

● The product is 2-bromopropane (Structure **3.2**).

$$\overset{I}{\underset{|}{CH_3CHCH_3}}$$

3.1 🖳

$$Br^- + \overset{I}{\underset{|}{CH_3CHCH_3}} \quad \longrightarrow \quad \overset{Br}{\underset{|}{CH_3CHCH_3}} \quad + \quad I^-$$

$$\textbf{3.1} \qquad\qquad\qquad \textbf{3.2}$$

🖳

This type of reaction is known as a **nucleophilic substitution reaction**, often abbreviated to S_N **reaction**. As we shall see in the next section, it is the nature of the leaving group X that is primarily responsible for the success or failure of most S_N reactions. The leaving group X serves two roles in an S_N reaction. It must form a polarized covalent bond to carbon in the substrate in order to provide the electrophilic site, but, more importantly, it must become a stable entity when it departs with a pair of electrons. For both these reasons, hydrogen atoms are not substituted in S_N reactions. To act as a leaving group, a hydrogen atom must gain a second electron, thus forming a hydride anion (H^-). However, the hydride anion is a very high-energy species. In the next section, we shall discuss the types of nucleophile and leaving group that can be employed in S_N reactions.

Let's first summarize the main features of the reaction in Equation 3.1:

nucleophilic substitution reaction

$$Nu^- + \quad -\overset{|}{\underset{|}{C}}-X \quad \longrightarrow \quad -\overset{|}{\underset{|}{C}}-Nu \quad + \quad X^- \qquad (3.1)$$

nucleophile   electrophilic carbon centre   leaving group departs as stable entity

* In this book, we follow the convention of using the symbol Nu^- to represent a charged nucleophile, and Nu: to represent an uncharged nucleophile (uncharged nucleophiles are discussed in Section 3.2.1).

COMPUTER ACTIVITY 3.1 Curly arrows

At some point in the near future you should study *Curly arrows* in the CD-ROM that accompanies this book. This activity reviews the material covered so far in this book and introduces you to curly arrows, a powerful method for keeping track of electrons in bond-making and bond-breaking processes. Curly arrows are used extensively, so it is important that you have a clear understanding of their meaning. If you cannot study the CD-ROM immediately, don't worry, you can read on since the text in the next few sections does not rely on the new concepts taught on the CD-ROM.

QUESTION 3.1

Identify the nucleophile, electrophilic carbon atom, spectator ion and leaving group in each of the following S_N reactions.

(i) CH_3CH_2Cl + $Na^{+-}SCH_3$ $\longrightarrow$ $CH_3CH_2SCH_3$ + $Na^+ Cl^-$

(ii)

CH_2Cl (benzene ring) + $K^{+-}CN$ $\longrightarrow$ CH_2CN (benzene ring) + $K^+ Cl^-$

(iii)

H Br (cyclohexane ring) + $Na^+ I^-$ $\longrightarrow$ H I (cyclohexane ring) + $Na^+ Br^-$

Each of the organic reactants and products in (i) to (iii) above is available in WebLab ViewerLite on the CD-ROM associated with this book.

3.2 The scope of the S_N reaction

So far, you have seen that the requirements for an S_N reaction are (i) a nucleophile and (ii) a substrate containing a suitable leaving group. The S_N reaction is extremely versatile, and a large number of combinations of nucleophile and leaving group can be employed. Let's discuss first the range of nucleophiles and leaving groups, and then the practicalities of designing and carrying out a nucleophilic substitution reaction.

3.2.1 Nucleophiles

Nucleophiles were introduced in Section 3.1 and can be formally defined as follows:

> All nucleophiles have at least one non-bonded pair of electrons, which ultimately form a new covalent bond.

A list of nucleophiles commonly employed in S_N reactions is given in Table 3.2, together with examples of typical reactions involving these nucleophiles. Look at this list, and check that all the nucleophiles shown are encompassed by the above definition.

Table 3.2 Nucleophiles commonly employed in S_N reactions and some typical examples for each

		Nucleophile	Reaction
(a)	Halides	Cl^-	$RX + Cl^- \longrightarrow RCl + X^-$
		Br^-	$RX + Br^- \longrightarrow RBr + X^-$
		I^-	$RX + I^- \longrightarrow RI + X^-$
(b)	Oxygen or sulfur as nucleophilic atom	$^-OH, {}^-SH$	$RX + {}^-OH \longrightarrow ROH + X^-$
		RO^-, RS^-	$RX + {}^-SCH_3 \longrightarrow RSCH_3 + X^-$
		$H_2O, H_2S, ROH, RSH,$	$RX + H_2O \longrightarrow ROH + HX$
(c)	Nitrogen as nucleophilic atom	NH_3	$RX + NH_3 \longrightarrow R\overset{+}{N}H_3X^-$
		RNH_2, R_2NH, R_3N	$RX + (CH_3)_3N \longrightarrow (CH_3)_3\overset{+}{N}RX^-$
(d)	Carbon as nucleophilic atom	^-CN	$RX + {}^-CN \longrightarrow RCN + X^-$
		$RC{\equiv}C^-$	$RX + Na^+ \, {}^-C{\equiv}CCH_3 \longrightarrow RC{\equiv}CCH_3 + Na^+X^-$

In Table 3.2, R = benzyl or alkyl, e.g. methyl or ethyl; X = leaving group.

The nucleophiles in the third row of (b) are useful only in special circumstances, as you will see in Section 3.2.2.

Note also that $RC{\equiv}C^-$ in the lower row of (d) is prepared by treating the corresponding alkyne, $RC{\equiv}CH$ with sodium amide $Na^+ \, {}^-NH_2$.

Some of the nucleophiles in Table 3.2 should now be familiar to you, and the majority of them are anionic (that is, they bear a negative charge). However, you may be surprised to see neutral molecules, such as ammonia and the amines included as well.

○ Do ammonia and the amines satisfy the definition of a nucleophile?

○ Even though ammonia and the amines are neutral, they do have a non-bonded pair of electrons available, on the nitrogen atom, to form a covalent bond.

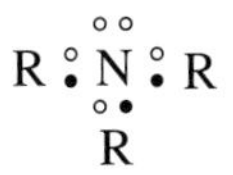

$$R \overset{\circ\circ}{\underset{\underset{R}{\circ\,\bullet}}{\,:\,N\,:\,}} R$$

R stands for H in ammonia and for alkyl or aryl groups in the amines.

Nitrogen is in Group V of the Periodic Table, and so has five outer-shell electrons. Three of these electrons participate in covalent bonding; the other two form a non-bonded electron pair, which enables nitrogen compounds (such as ammonia, Figure 3.3 overleaf) to act as nucleophiles in S_N reactions. This is illustrated by Reaction 3.2, between ammonia and iodomethane, giving methylammonium iodide.

$$(3.2)$$

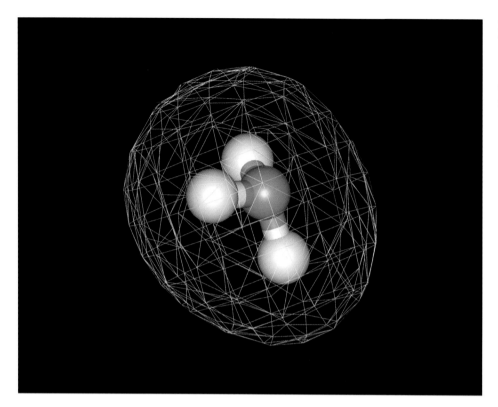

Figure 3.3
The charge density distribution in ammonia. The red areas of the surface indicate regions of high electron density and the blue areas indicate low electron density. 💻

Note that the tetravalent nitrogen atom bears a positive charge. When its non-bonded electron pair forms a bond, the nitrogen atom loses control of one electron, losing it to the carbon atom of iodomethane. The nitrogen thus becomes positively charged.

Neutral molecules containing the hydroxyl group are seldom employed as nucleophiles, even though the oxygen atom has two non-bonded pairs of electrons. We shall see why this is so later, but, in general, the anions derived from water ($^-$OH), alcohols (RO$^-$) and carboxylic acids (RCOO$^-$) are used in preference to the parent compounds. Anions are usually employed in the synthesis of ethers and ethanoates (acetates) from haloalkanes (Reactions 3.3 and 3.4)

$$RBr \ + \ Na^{+ \ -}OCH_3 \ \longrightarrow \ ROCH_3 \ + \ Na^+ \ Br^- \qquad (3.3)$$

$$RBr \ + \ Na^{+ \ -}O\overset{\overset{O}{\parallel}}{C}CH_3 \ \longrightarrow \ RO\overset{\overset{O}{\parallel}}{C}CH_3 \ + \ Na^+ \ Br^- \qquad (3.4)$$

Having seen the variety of nucleophiles that can be employed in S_N reactions, let's now go on to look at leaving groups.

3.2.2 Leaving groups

⬤ What are the main criteria that a leaving group (X) must fulfil in an S_N reaction such as Reaction 3.1?

⬤ The leaving group must (i) polarize the covalent bond such that the attached carbon is electrophilic and (ii) become a relatively stable entity when it departs with both electrons.

The leaving groups most commonly encountered in S_N reactions are shown in Table 3.3. Look at this table, and see if the leaving groups listed there fulfil the above definition and fit in with the reactions that you have seen so far in this book.

Table 3.3 Leaving groups (X) commonly employed in S_N reactions

Substrate(RX)^a	X as product
RI	I$^-$
RBr	Br$^-$
RCl	Cl$^-$

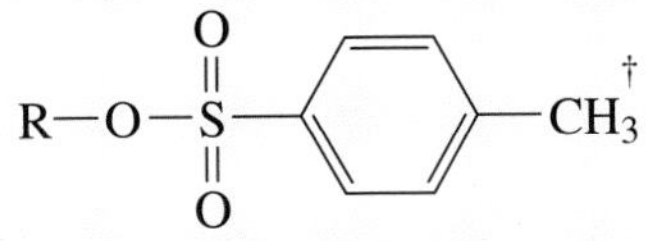

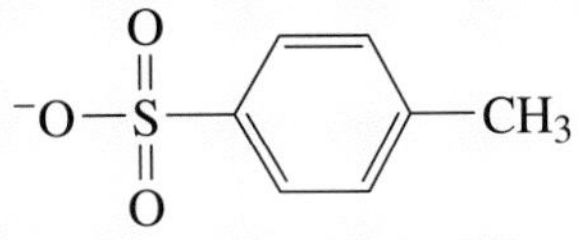

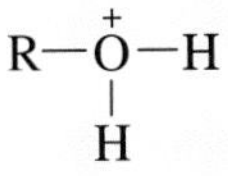

^{a}R, R^1 and R^2 are alkyl groups. The structure marked † is sometimes abbreviated to R—OTs, short for tosylate.

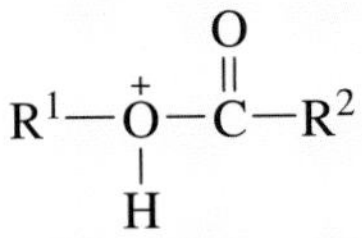

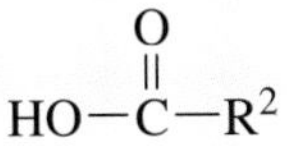

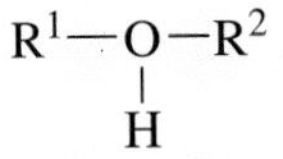

The leaving groups most commonly employed in S_N reactions are the halogens, since halide anions are particularly stable because they have an octet of valence electrons.

In contrast to the halogen leaving groups, the oxygen-linked leaving groups are rather difficult to understand. Hydroxyl (−OH) does not feature as a leaving group (to form $^-$OH) in Table 3.3, and you have already seen that ethanol does not react with sodium bromide. It is found that S_N reactions involving hydroxyl as the leaving group are rarely, if ever, successful. Hydroxyl is said to be a poor leaving group. The same is true of alkoxyl (RO−) and carboxyl groups

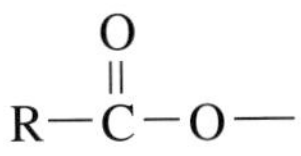

One rationalization of these observations correlates the leaving group ability of X (in RX) with the acidity of the corresponding acid HX. The stronger the acid, the larger the value of K_a, the acid dissociation constant for HX

$$HX \;\; \underset{}{\overset{K_a}{\rightleftharpoons}} \;\; H^+ \; + \; X^-$$

You can see how good a correlation this is, by looking at the acid dissociation constant data in Table 3.4. The pHs of aqueous solutions of three of the substances in Table 3.4 (overleaf) are shown experimentally in Figure 3.4 (overleaf).

Table 3.4 Some acid dissociation constants with substances HX ranked in descending order of ability of the leaving group X

Substance	Acid dissociation constant K_a/mol dm^{-3}
HI	10^{10}
HBr	10^9
HCl	10^7
$R^1-\overset{+}{\underset{H}{O}}-\overset{\overset{O}{\parallel}}{C}-R^2$	10^7
HO$-\overset{\overset{O}{\parallel}}{\underset{\underset{O}{\parallel}}{S}}$—⬡—CH$_3$	10^7
$R\overset{+}{O}H_2$	10^2
H_3O^+	10^2
HO$-\overset{\overset{O}{\parallel}}{C}-CH_3$	10^{-5}
HCN	10^{-10}
HOCH$_2$CH$_3$	10^{-16}
HOH	10^{-16}

⬤ Look at Table 3.4. What is the relationship between the leaving group ability of X and the acidity of HX?

⬤ The acid (HX) corresponding to a good leaving group (X) has a large acid dissociation constant K_a.

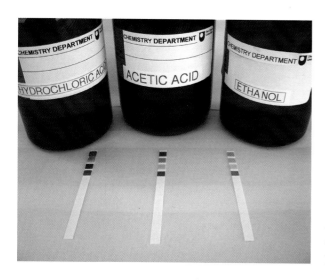

Figure 3.4
pH papers can be used to measure the pH of an aqueous solution of hydrochloric acid, ethanoic acid (acetic acid) and ethanol. The results are pH 1, pH 4 and pH 7 respectively.

Put another way, good leaving groups give rise to a stable entity, for example, X^-, and if X^- is stable, HX will be a strong acid.

Experimentally, the halogens are good leaving groups, and the acid dissociation constants of HI, HBr and HCl are all very large ($>10^7$). Experimentally, hydroxyl, −OH, is found to be a poor leaving group. The corresponding acid (HOH or H_2O) has an acid dissociation constant of 10^{-16}. So the correlation of acidity and leaving group ability appears to stand up well.

We saw in Section 2.1 that alcohols do undergo S_N reactions when treated with acids such as HBr. In these instances, the hydroxyl group is converted into a good leaving group before the S_N reaction takes place. So, in the reaction between ethanol and hydrogen bromide, protonation of the oxygen atom in ethanol occurs in the first step and the resulting trivalent oxygen atom bears a positive charge. The substitution occurs during the second step (Reaction 3.5). So, in this process the leaving group in the S_N reaction is the water molecule, which certainly fulfils the criterion that a leaving group should be a stable entity.

$$CH_3CH_2\overset{\bullet\bullet}{\underset{\bullet\bullet}{O}}H \;+\; H^+Br^- \longrightarrow CH_3CH_2 - \overset{+}{\underset{H}{\overset{H}{O}}}\colon \;+\; Br^- \xrightarrow[\text{step}]{\text{substitution}} CH_3CH_2Br \;+\; H_2\overset{\bullet\bullet}{\underset{\bullet\bullet}{O}} \qquad 3.5)$$

Does this observation accord with the suggestion that the protonated form of a good leaving group has a large dissociation constant?

Yes. The protonated form of water (H_3O^+) has a dissociation constant of 10^2 which is well over half-way up the logarithmic scale outlined in Table 3.4.

Similar arguments apply to S_N reactions involving ethers and esters. RO− and R−COO− are poor leaving groups (ROH and R−COOH are weak acids), but in acidic media, protonation occurs, and S_N reactions then become possible.

$$CH_3OCH_3 \;+\; HI \longrightarrow \underset{\text{a strong acid}}{H_3C - \overset{+}{\underset{H}{O}} - CH_3} \;+\; I^- \longrightarrow CH_3I \;+\; CH_3OH$$

$$\underset{}{CH_3CH_2\overset{\overset{\displaystyle O}{\parallel}}{O}CCH_3} \;+\; HBr \longrightarrow \underset{\text{a strong acid}}{CH_3CH_2 - \overset{+}{\underset{H}{O}} - \overset{\overset{\displaystyle O}{\parallel}}{C} - CH_3} \;+\; Br^- \longrightarrow CH_3CH_2Br \;+\; CH_3\overset{\overset{\displaystyle O}{\parallel}}{C}OH$$

So, one of the problems posed earlier can now be understood. Alcohols are not converted into bromoalkanes by sodium bromide because hydroxyl is a poor leaving group

$$ROH \;+\; Na^+\,Br^- \;\;\xrightarrow{\quad\times\quad}\;\; RBr \;+\; Na^+\,{}^-OH \qquad (\text{the } K_a \text{ of } H_2O \text{ is } 10^{-16})$$

Under acidic conditions, with hydrogen bromide, the leaving group is water, so the reaction proceeds readily

$$ROH \;+\; HBr \longrightarrow R\overset{+}{\underset{H}{-O-H}} \;+\; Br^- \longrightarrow RBr \;+\; H_2O \quad \text{(the } K_a \text{ of } H_3O^+ \text{ is } 10^2\text{)}$$

Similarly, alcohols can be converted into chloroalkanes by treatment with HCl

$$ROH \;+\; HCl \longrightarrow R\overset{+}{\underset{H}{-O-H}} \;+\; Cl^- \longrightarrow RCl \;+\; H_2O$$

The *para*-toluenesulfonate ester group (often abbreviated to tosylate, or OTs and shown in Figure 3.5) is also a good leaving group (the acid dissociation constant of $CH_3C_6H_4SO_3H$ is 10^7), and so S_N reactions are possible (for example, Reaction 3.6)

$$K^+\ {}^-CN \;+\; RCH_2-OSO_2-\!\!\left\langle\!\!\bigcirc\!\!\right\rangle\!\!-CH_3 \longrightarrow RCH_2CN \;+\; K^+\,{}^-OSO_2-\!\!\left\langle\!\!\bigcirc\!\!\right\rangle\!\!-CH_3 \qquad (3.6)$$

It is fairly easy to convert an alcohol into a tosylate by treating it with *para*-toluenesulfonyl chloride, so this would be an alternative way of converting the poor OH leaving group into a good leaving group.

$$RCH_2-OH \;+\; Cl-SO_2-\!\!\left\langle\!\!\bigcirc\!\!\right\rangle\!\!-CH_3 \longrightarrow RCH_2-OSO_2-\!\!\left\langle\!\!\bigcirc\!\!\right\rangle\!\!-CH_3 \;+\; HCl$$

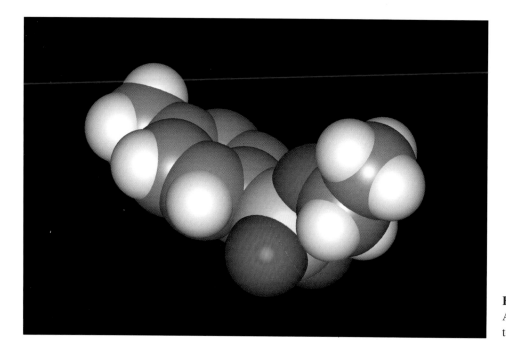

Figure 3.5
A space-filling model of a *para*-toluenesulfonate ester.

Another point worth emphasizing here is that C—H and C—CR$_3$ systems do not undergo S$_N$ reactions to replace H and CR$_3$ by a nucleophile.

◯ What leaving groups would be generated if such S$_N$ reactions did occur?

◯ H$^-$ and $^-$CR$_3$. These are very unlikely leaving groups, because the acid dissociation constants of H$_2$ and HCR$_3$ are very small (less than 10^{-45}).

So there are very few leaving groups that you have to remember. There are the halides and tosylates, together with alcohols, ethers, and carboxylates, where the last three groups undergo S$_N$ reactions only under acidic conditions. This classification only refers to substitution at a *saturated* carbon atom. When substitution occurs at a carbonyl carbon atom, a number of groups that we have called poor leaving groups here, are perfectly adequate leaving groups.

STUDY NOTE

By now you should have an idea of what constitutes a good nucleophile and a good leaving group. You should familiarize yourself with the lists of common nucleophiles and leaving groups given in Tables 3.2 and 3.3.

QUESTION 3.2

Predict the products of the following reactions, all of which proceed by an S$_N$ mechanism. Write a balanced equation for each reaction.

(i) 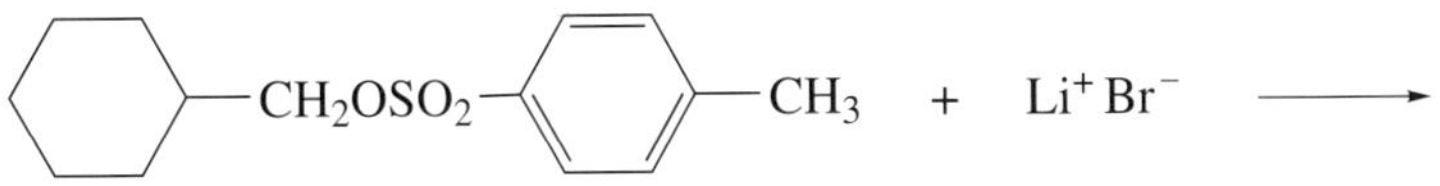 ⬡—CH$_2$OSO$_2$—⬡—CH$_3$ + Li$^+$Br$^-$ ⟶

(ii) C$_6$H$_5$CH$_2$Br + Na$^+$ $^-$SH ⟶

(iii) $\overset{..}{N}$H$_3$ + CH$_3$I ⟶

(iv) CH$_3$I + CH$_3$C≡C$^-$Na$^+$ ⟶

Each of the organic reactants and products in (i) to (iv) above is available in WebLab ViewerLite on the CD-ROM associated with this book.

3.3 How far and how fast?

In our discussion of nucleophilic substitutions we have looked at several S$_N$ reactions, and we have identified the structural features (nucleophile, leaving group, electrophilic carbon atom) essential to the success of these reactions. But in order to appreciate the full scope of the S$_N$ reaction we need to be able to predict the outcome of combining a nucleophile from Table 3.2 with a substrate containing a leaving group from Table 3.3. We need to have some idea of

• whether the reaction is thermodynamically feasible, that is, will it go?

• whether the reaction is kinetically feasible, that is, will it go in a reasonable time?

3.3.1 How far?

Any reaction can in principle be thought of as an equilibrium, so let's think of the S$_N$ reaction in this way. This is reasonable because the leaving group (once it has left as X$^-$) has electrons available for bonding, and so in theory it could act as a nucleophile in the reverse reaction. Indeed, the halide anions feature in both Tables 3.2 and 3.3: they can act as both nucleophiles and leaving groups. So we can write

RX + Nu$^-$ ⇌ RNu + X$^-$

● Write out the expression for the equilibrium constant for this reaction.

● $K = \dfrac{[\text{RNu}][\text{X}^-]}{[\text{RX}][\text{Nu}^-]}$

● Which thermodynamic quantity is directly related to the equilibrium constant? What is the numerical relationship between this thermodynamic quantity and the equilibrium constant?

● It is $\Delta G_m^\ominus$, the standard molar Gibbs function change, that is directly related to the equilibrium constant. The two terms are related via the equation

$$\Delta G_m^\ominus = -2.303RT \log K$$

where R is the gas constant and T is the absolute temperature.

$\Delta G_m^\ominus$ can be calculated from standard thermodynamic data. Unfortunately, it isn't possible to calculate the values of $\Delta G_m^\ominus$ for the majority of substitution reactions, because the standard data in solution aren't available. So how can we predict the outcome of a given S_N reaction?

The answer lies in the judicious use of experimental results. Once information is obtained concerning nucleophiles and leaving groups that have been successfully employed in S_N reactions, it becomes possible to predict with a fair degree of accuracy which reactions will give high yields of products. All the reactions discussed in this book involve good nucleophiles and good leaving groups, as shown in Tables 3.2 and 3.3. A reaction involving a nucleophile from Table 3.2 and a leaving group from Table 3.3 would be expected to give a good yield of substitution product (that is, K would be large). However, if the leaving group is also a good nucleophile (that is, if the leaving group appears in *both* Tables 3.2 and 3.3), a mixture of reactants and products will generally be obtained and the value of K will be about 1. For example, halogen exchange reactions usually result in only partial reaction, because halide anions can act as both nucleophiles and leaving groups.

$$\text{RCl} + \text{Na}^+\text{I}^- \rightleftharpoons \text{RI} + \text{Na}^+\text{Cl}^-$$

At first sight this reaction does not appear to be useful from a preparative viewpoint; no one wants to have to separate products from unchanged starting materials unless it's absolutely necessary.

● Think about Le Chatelier's principle. Is there a way of ensuring that the iodoalkane (RI) is the major reaction product?

● If the concentration of one of the reactants were increased, or if one of the products were removed, then according to Le Chatelier's principle, reaction would take place to re-establish equilibrium. In this example, if the concentration of sodium iodide were increased so that an excess were present, the equilibrium would re-establish itself by producing more products. Similarly, removal of sodium chloride, or the iodoalkane, would also shift the equilibrium to the right.

The general procedure for converting chloroalkanes into iodoalkanes uses both of these techniques in order to ensure a good yield of the product. The chloroalkane and an excess of sodium iodide are usually dissolved in propanone (acetone), and the mixture is boiled. Unlike sodium iodide, sodium chloride is insoluble in propanone, and so it is precipitated as it is formed, and thus removed from the equilibrium. This serves to force the reaction in the desired direction.

So when the nucleophile is also a good leaving group, and the leaving group is also a good nucleophile, experimental conditions can generally be found to facilitate the desired reaction.

BOX 3.1 Another difference between Organic and Inorganic Chemistry

As you are no doubt aware by now, Chemistry is divided up into various subdisciplines: mainly organic, inorganic and physical chemistry. Such divisions were originally historical, but because of the nature of carbon compounds compared to those of other elements, the division does make some sense. For example, in this book you will see how, by changing the conditions of a reaction, you can easily swap from one product predominating to another product being formed instead. This is because the Gibbs energy changes of organic reactions tend to be quite small compared to those of inorganic reactions, and the equilibrium constants are therefore not that big and are more susceptible to changes in the conditions.

3.3.2 How fast?

The rate of a reaction is a measure of how fast the reactants are converted into the products. In this section we want to develop a qualitative idea of how the rate of S_N reactions depends on the nature of the nucleophile. This can best be illustrated by looking at the reaction of one organic substrate, bromomethane, with a variety of nucleophiles under similar conditions (Reaction 3.7)

$$Nu^- + CH_3Br \rightleftharpoons CH_3Nu + Br^- \tag{3.7}$$

A comparison of the reaction rates should then indicate the influence of the nucleophile on the rate of reaction; such data are given in Table 3.5.

By carrying out similar experiments, it is possible to draw up a list of nucleophiles in order of their reactivity or **nucleophilicity**. The more reactive, or nucleophilic, the nucleophile, the faster the reaction under standard conditions. Encounters between substrate molecules and good (that is, more reactive) nucleophiles are much more likely to result in reaction than encounters between substrate molecules and poor (that is, less reactive) nucleophiles.

A list of nucleophiles, in approximate order of reactivity, is given in the box below. Species separated by a comma in the list have reactivities of the same order of magnitude.

most reactive ^-CN, I^-, ^-SH, $RS^- > {}^-OH$, $RO^- > Br^- > Cl^- > CH_3COO^- >$ $F^- > NH_3$, $RSH \gg H_2O$, ROH, RSO_3^- *least reactive*

This order is only approximate, because nucleophilicities are very solvent-dependent, and varying the reaction solvent may lead to a change in the order.

Table 3.5 Relative rates of reaction of bromomethane with various nucleophiles in water at 50 °C

Nucleophile	Relative rate
H_2O	1
CH_3COO^-	5.2×10^2
Cl^-	1.1×10^3
Br^-	7.8×10^3
^-OH	1.6×10^4
I^-	1.1×10^5

The following general points are worth noting:

1   Anions are generally better nucleophiles than the corresponding neutral molecules. This is especially true when comparing $^-$OH, RO$^-$ and RS$^-$ with their protonated forms.

2   Within a Period or row of the Periodic Table, the better nucleophile has the lower atomic number. This means that nitrogen nucleophiles such as ammonia and amines are better nucleophiles than the corresponding oxygen nucleophiles, water and alcohols. In fact, water and alcohols generally react too slowly to be of use, and often it is necessary to use their anionic forms HO$^-$ or RO$^-$.

3   Within a Group in the Periodic Table, nucleophilicity generally increases with increasing atomic number. This is shown by the following orders of reactivity: I$^-$ > Br$^-$ > Cl$^-$ > F$^-$; and RS$^-$ > RO$^-$.

Having looked at the way in which the rates of S_N reactions vary with different nucleophiles, in the next section we look at the influence of the substrate on the rate of reaction. Some very interesting mechanistic conclusions result from this discussion.

3.4 Summary of Section 3

1   Nucleophilic substitution (S_N) reactions are of the following type

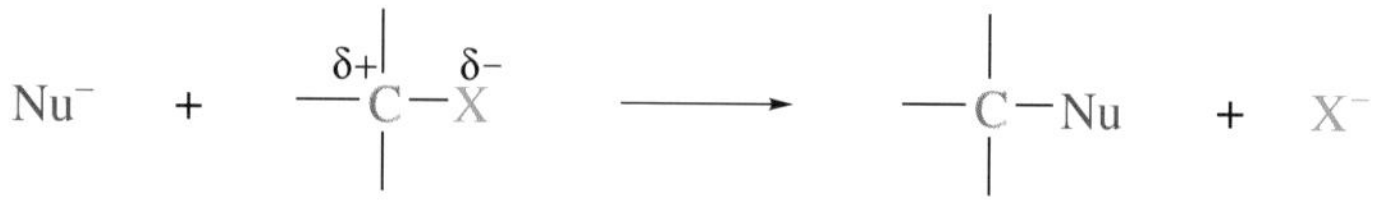

$$Nu^- \;+\; \overset{\displaystyle |}{\underset{\displaystyle |}{\overset{\delta+}{C}}}\!\!-\!\!\overset{\delta-}{X} \;\longrightarrow\; \overset{\displaystyle |}{\underset{\displaystyle |}{C}}\!\!-\!\!Nu \;+\; X^-$$

Group X, which was bonded to an aliphatic carbon atom, is replaced by a new group Nu. Nu$^-$ is called the nucleophile. It must have a non-bonded pair of electrons to form a new covalent bond. The group X is called the leaving group. It forms a polarized bond to carbon, in which the carbon atom is slightly positive ($C^{\delta+}$). X must form a stable entity, X$^-$, when it departs. The carbon atom bearing a slight positive charge is said to be electrophilic; it attracts the nucleophile.

2   Nucleophiles generally bear a negative charge (for example, $^-$OH, $^-$CN, I$^-$), but ammonia and amines, and in some cases water and alcohols, can also act as nucleophiles.

3   The halides and *para*-toluenesulfonate are the most commonly employed leaving groups. Hydroxyl groups, ethers and esters act as leaving groups only when they are protonated. The leaving group ability of a group (X) can be gauged by the acid dissociation constant K_a, of the corresponding acid (HX). Carbon–hydrogen and carbon–carbon bonds are rarely broken in S_N reactions.

4   Good yields of substitution products will be obtained if a good nucleophile and a good leaving group are employed, and if the leaving group is not also a good nucleophile. When the leaving group is also a good nucleophile (for example, a halide) a mixture of reactants and products will generally be formed. Good yields of the products can still be obtained from such equilibria by exploiting Le Chatelier's principle.

5   For a given substrate, the rate of reaction depends on the reactivity (nucleophilicity) of the nucleophile. The following approximate order generally applies, although this order is solvent-dependent

$^-$CN, I$^-$, $^-$SH, RS$^-$ > $^-$OH, RO$^-$ > Br$^-$ > Cl$^-$ > CH$_3$COO$^-$ > F$^-$ > NH$_3$, RSH $\gg$ H$_2$O, ROH, RSO$_3$$^-$.

Notice that (i) anions are generally better nucleophiles than the corresponding neutral molecules, (ii) within a row of the Periodic Table, the better nucleophile has the lower atomic number, (iii) within a Group in the Periodic Table, nucleophilicity increases with increasing atomic number.

QUESTION 3.3

Would you expect the following S_N reactions to proceed to give the products shown, not proceed at all, or lead to a mixture of reactants and products?

(i) $\quad CH_3CN \; + \; Na^+I^- \quad \longrightarrow \quad CH_3I \; + \; Na^{+-}CN$

(ii) $\quad CH_3CH_2OSO_2$—⬡—$CH_3 \; + \; Na^{+-}SCH_3 \quad \longrightarrow$

$Na^{+-}OSO_2$—⬡—$CH_3 \; + \; CH_3CH_2SCH_3$

(iii) $\quad C_6H_5CH_2OH \; + \; Na^+Br^- \quad \longrightarrow \quad C_6H_5CH_2Br \; + \; Na^{+-}OH$

(iv)

CH_2Cl–⬡ $\; + \; Na^+I^- \quad \longrightarrow \quad$ CH_2I–⬡ $\; + \; Na^+Cl^-$

(v) $\quad CH_3CH_2OH \; + \; HCl \quad \longrightarrow \quad CH_3CH_2Cl \; + \; H_2O$

Each of the organic reactants and products in (i) to (v) above is available in WebLab ViewerLite on the CD-ROM associated with this book.

QUESTION 3.4

State which reaction of each of the following pairs should go faster and why. If a decision isn't possible, say so.

(i) (a) $\quad CH_3CH_2OSO_2$—⬡—$CH_3 \; + \; Na^+I^- \quad \longrightarrow \quad Na^{+-}OSO_2$—⬡—$CH_3 \; + \; CH_3CH_2I$

(b) $\quad CH_3CH_2OSO_2$—⬡—$CH_3 \; + \; Na^+Cl^- \quad \longrightarrow \quad Na^{+-}OSO_2$—⬡—$CH_3 \; + \; CH_3CH_2Cl$

(ii) (a) $\quad CH_3CH_2Br \; + \; Na^{+-}OH \quad \longrightarrow \quad CH_3CH_2OH \; + \; Na^+Br^-$

(b) $\quad CH_3CH_2Br \; + \; H_2O \quad \longrightarrow \quad CH_3CH_2OH \; + \; HBr$

(iii)(a) $\quad CH_3CH_2Br \; + \; Na^{+-}CN \quad \longrightarrow \quad CH_3CH_2CN \; + \; Na^+Br^-$

(b) $\quad (CH_3)_2CHBr \; + \; Na^{+-}CN \quad \longrightarrow \quad (CH_3)_2CHCN \; + \; Na^+Br^-$

Each of the organic reactants and products in (i) to (iii) above is available in WebLab ViewerLite on the CD-ROM associated with this book.

S_N2 AND S_N1 REACTION MECHANISMS

4

4.1 Introduction

So far, you have seen that nucleophilic substitution reactions proceed if a compound with an electrophilic carbon atom attached to a good leaving group is treated with a nucleophilic reagent under the appropriate reaction conditions. You might think that, given a knowledge of the available nucleophiles and leaving groups, this is sufficient information to be able to use S_N reactions in synthesis.

But the ultimate aim of anyone planning to use an S_N reaction is to be able to predict the outcome of the reaction with certainty, and there are still several points that we have to clarify before we can do this. For example, we now have some idea about the way in which the rate of a reaction varies with different nucleophiles, but as yet we haven't discussed how the rate varies with different substrate structures. Table 4.1 shows some qualitative rate data for the conversion of two bromoalkanes into the corresponding ethyl ethers (ethoxyalkanes) using ethanol and sodium ethoxide as nucleophiles. The reaction is said to be complete when greater than 99% of the substrate is converted into product. Look at the information in Table 4.1.

● Does the fact that bromoethane undergoes substitution faster with the ethoxide anion than with ethanol fit with the order of reactivity of nucleophiles?

● Yes, the order tells us that ethoxide ion (^-OR) is a much more reactive nucleophile than ethanol (ROH).

Table 4.1 Rate data for reactions $RBr \longrightarrow ROCH_2CH_3$

Substrate, RBr	Conditions	Rate
CH_3CH_2Br	CH_3CH_2OH/boil	50% complete in 4 days
CH_3CH_2Br	$NaOCH_2CH_3$/solvent, CH_3CH_2OH	complete in a few minutes
$(CH_3)_3CBr$	CH_3CH_2OH/boil	complete in a few minutes

The surprising information from Table 4.1 is that, with ethanol, the compound 2-bromo-2-methylpropane undergoes complete reaction in a matter of minutes, compared with the days needed for bromoethane. Observations such as these made organic chemists think in more detail about the mechanism of the S_N reaction. This subject was of particular interest to two organic chemists working at University College, London in the 1930s — Professor C. K. Ingold (Figure 4.1) and Professor E. D. Hughes (Figure 4.2). They carried out a series of S_N reactions, and measured the way in which the reaction rates varied with changing concentrations of the reactants. They then tried to explain their results on a molecular level, in terms of the mechanism, that is, the order in which bonds are made and broken during S_N reactions. Many of the puzzling observations concerning the S_N reaction were explained by Ingold and Hughes, who through their pioneering work ushered in the era of mechanistic organic chemistry.

Christopher Kelk Ingold (1893–1970)

Christopher Ingold was born in 1893. He attended Hartley College, Southampton where he obtained a second-class degree (which he attributed to playing too much chess). He then went to Imperial College as a lecturer and then to Leeds University as Professor of Organic Chemistry. He returned to London in 1930 to University College, where he stayed until he retired. He was knighted in 1958. He was one of the founders of mechanistic organic chemistry and was responsible for developing the electronic theory of organic reactions, including ideas such as nucleophiles, electrophiles, inductive effects and resonance effects. With Cahn and Prelog he devised the priority rules for defining stereochemistry.

Figure 4.1
Sir Christopher
Ingold (1893–1970).

Edward David Hughes (1906–1963)

Hughes graduated from University of Wales, Bangor and joined University College London at the same time as Ingold. They collaborated for the next forty years on mechanistic organic chemistry. As well as elucidating the mechanisms of substitution and elimination, he pioneered the use of isotopes for labelling compounds in studies of reaction mechanisms. He is alleged to have tried to throw a weighing machine into the sea at Aberystwyth for telling him he weighed more than 16 stones. He and other students were prevented from consigning the machine to the waves by the appearance of a policeman!

Figure 4.2
Edward Hughes
(1906–1963).

4.2 Kinetics and mechanism of S$_N$ reactions

Part 1 of this book explained how information about reaction mechanisms can be gained from a study of reaction rates. In this section, we shall use such kinetic evidence to help unravel the mechanism of substitution reactions. The S$_N$ reaction between bromomethane and sodium hydroxide in methanol at 25 °C follows second-order kinetics, with the experimental rate equation

$$J = k[CH_3Br][NaOH] \tag{4.1}$$

However, the S_N reaction between 2-bromo-2-methylpropane and sodium hydroxide in methanol at 25 °C follows first-order kinetics, with the experimental rate equation

$$J = k[(CH_3)_2CHBr] \qquad\qquad (4.2)$$

In fact, analysis of many different substitution reactions reveals only two types of rate equations for substitution at a carbon atom

either $J = k[\text{substrate}][\text{nucleophile}]$

or $J = k[\text{substrate}]$

We can use this information to determine two possible mechanisms of substitution.

The overall substitution reaction involves replacing a leaving group by a nucleophile, and there are three possible ways in which this can happen:

(i)   The bond to the nucleophile forms *at the same time* as the bond to the leaving group breaks; a concerted mechanism (Section 4.2.1).

(ii)   The bond to the nucleophile forms *before* the bond to the leaving group breaks; a two-step associative mechanism (Section 4.2.2).

(iii)   The bond to the nucleophile forms *after* the bond to the leaving group breaks; a two-step dissociative mechanism (Section 4.2.3).

Let's analyse the kinetics of each of these mechanisms in detail.

4.2.1 A concerted mechanism

This is a one-step mechanism in which the incoming nucleophile forms a bond to the electrophilic carbon atom at the same time that the bond to the leaving group is broken

$$HO^- \; + \; CH_3{-}Br \quad\xrightarrow{\text{via }[\,HO\cdots CH_3\cdots Br\,]^-}\quad HO{-}CH_3 \; + \; Br^-$$

This elementary reaction has an activated complex, $[\,HO\cdots CH_3\cdots Br]^-$, in which the bond to the nucleophile is partially made and the bond to the leaving group is partially broken.

⬤ What is the theoretical rate equation predicted by this mechanism?

⬤ Since both substrate and reagent are involved in this elementary step the theoretical rate equation will involve a concentration term from both substrate and reagent

$$J = k[CH_3Br][^-OH]$$

4.2.2 Two-step associative mechanism

The bond between the nucleophile and the electrophilic carbon atom could form before the bond to the leaving group breaks, to give a discrete pentavalent reaction intermediate. Notice, in the previous mechanism (Section 4.2.1) the pentavalent species was an activated complex, but in this mechanism it is a reaction intermediate. The leaving group is then lost in the second step to give the expected product

Step 1

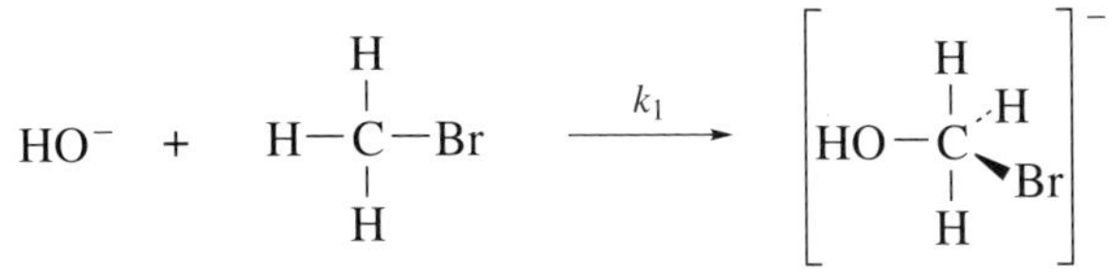

Step 2

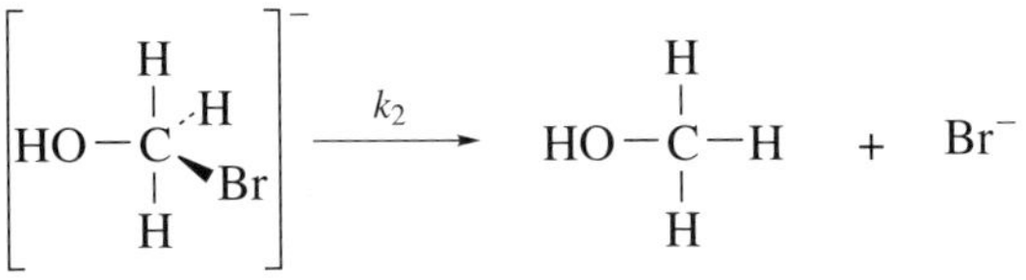

What is the theoretical rate equation predicted by this mechanism if (a) the first step is rate-limiting and (b) the second step is rate-limiting?

(a) If the first step is rate-limiting, the rate equation for this elementary reaction step will determine the predicted rate equation. Again, both substrate and reagent are involved in this elementary step and so the theoretical rate equation will involve a concentration term from both substrate and reagent

$$J = J_1 = k_1[CH_3Br][^-OH]$$

(b) If the second step is rate-limiting, the rate equation for this elementary reaction determines the predicted rate equation

$$J = J_2 = k_2[\mathbf{4.1}] \tag{4.3}$$

where [**4.1**] is the concentration of Structure **4.1**. To determine [**4.1**] we assume the first step is a rapidly established pre-equilibrium

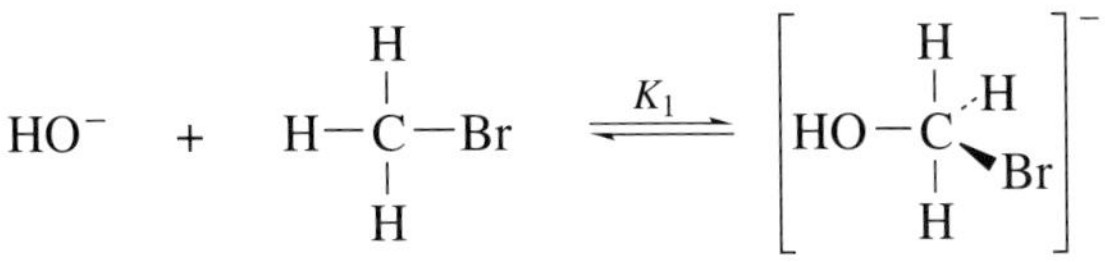

Thus we can write an expression for the equilibrium constant in Step 1

$$K_1 = \frac{[\mathbf{4.1}]}{[CH_3Br][^-OH]}$$

From this, [**4.1**] = $K_1[CH_3Br][^-OH]$, and substituting this into Equation 4.3 gives

$$J = k_2 K_1[CH_3Br][^-OH]$$

4.2.3 Two-step dissociative mechanism

The bond between the leaving group and the electrophilic carbon atom breaks before the bond to the nucleophile is formed, to give a discrete trivalent reaction intermediate. The nucleophile is then attached in the second step, to give the expected product

Step 1

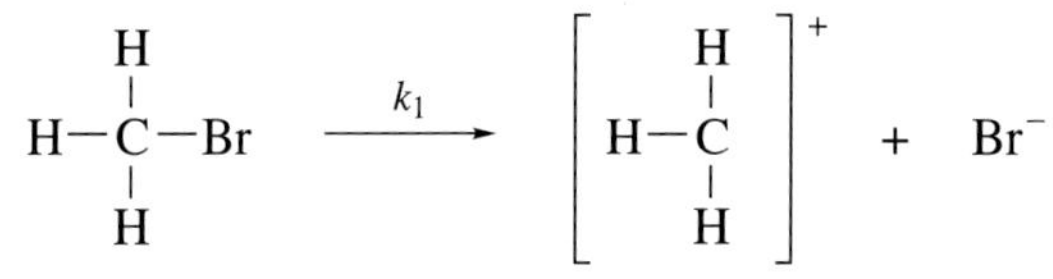

$$H-\overset{\overset{\displaystyle H}{|}}{\underset{\underset{\displaystyle H}{|}}{C}}-Br \xrightarrow{k_1} \left[H-\overset{\overset{\displaystyle H}{|}}{\underset{\underset{\displaystyle H}{|}}{C}} \right]^{+} + Br^{-}$$

4.2

Step 2

$$\left[H-\overset{\overset{\displaystyle H}{|}}{\underset{\underset{\displaystyle H}{|}}{C}} \right]^{+} + HO^{-} \xrightarrow{k_2} H-\overset{\overset{\displaystyle H}{|}}{\underset{\underset{\displaystyle H}{|}}{C}}-OH$$

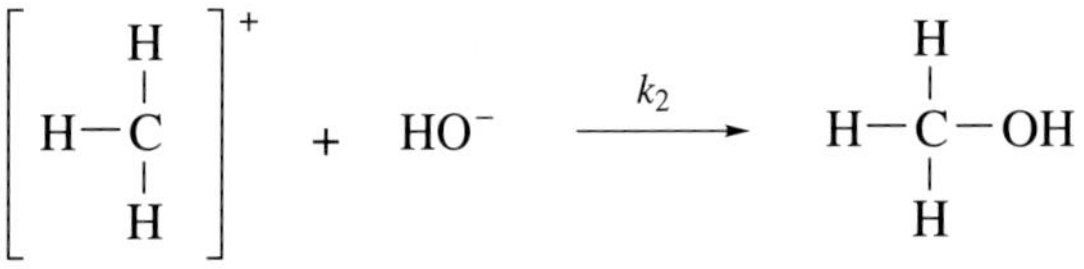

4.2

What is the theoretical rate equation predicted by this mechanism if (a) the first step is rate-limiting and (b) the second step is rate-limiting?

(a) If the first step is rate-limiting, the rate equation for this elementary reaction step will determine the predicted rate equation. Only the substrate is involved in this elementary step, giving the theoretical rate equation

$$J = J_1 = k_1[CH_3Br]$$

(b) If the second step is rate-limiting, the rate equation for this elementary reaction determines the predicted rate equation

$$J = J_2 = k_2[\mathbf{4.2}][^-OH] \qquad (4.4)$$

To determine [**4.2**] we assume the first step is a rapidly established pre-equilibrium

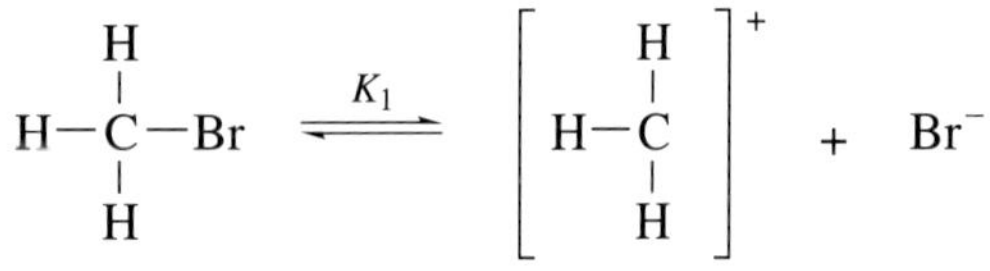

$$H-\overset{\overset{\displaystyle H}{|}}{\underset{\underset{\displaystyle H}{|}}{C}}-Br \underset{}{\overset{K_1}{\rightleftharpoons}} \left[H-\overset{\overset{\displaystyle H}{|}}{\underset{\underset{\displaystyle H}{|}}{C}} \right]^{+} + Br^{-}$$

4.2

Thus we can write an expression for the equilibrium constant in Step 1

$$K_1 = \frac{[\mathbf{4.2}][Br^-]}{[CH_3Br]}$$

From this

$$[\mathbf{4.2}] = \frac{K_1[CH_3Br]}{[Br^-]}$$

and substituting this into Equation 4.4 gives

$$J = \frac{k_2 K_1[CH_3Br][^-OH]}{[Br^-]}$$

4.2.4 Which mechanism is at work?

We have now determined the theoretical rate equations for each of our possible mechanisms. The next step is to compare them with experimental rate equations.

- The S$_N$ reaction between 2-bromo-2-methylpropane and sodium hydroxide in methanol at 25 °C follows first-order kinetics with the experimental rate Equation 4.2. Which is the most likely mechanism of this reaction?

- The only mechanism that generates a theoretical rate equation that is first-order is a two-step dissociative mechanism with the first step rate-limiting.

The S$_N$ reaction between bromomethane and sodium hydroxide in methanol at 25 °C follows second-order kinetics with the experimental rate Equation 4.1. The concerted and two-step associative mechanisms both generate a theoretical rate equation of this form, so it's at this point that we have to use other information to discriminate between these two mechanisms.

A trivalent carbon atom surrounded by six electrons, and bearing a positive charge (known as a **carbocation**, Figure 4.3) is relatively unstable. Such species have been observed spectroscopically and, in special cases can be isolated. However because the elements of the first row of the Periodic Table have small nuclei, they can only have eight valence electrons at most. A pentavalent carbon surrounded by ten electrons is thus extremely rare and we can discount the associative mechanism on chemical grounds — a concerted process is much more likely. As you will see in the CD-ROM, examination of the stereochemistry of the reactants and products of substitution confirms that the second-order substitutions proceed via the concerted mechanism.

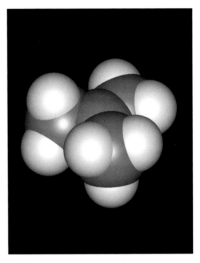

Figure 4.3
Space-filling model of a carbocation: notice the planar nature of the molecule.

Extensive analysis of rate data has identified two mechanisms for substitution at a carbon atom, and these have been given special names, based on the molecularity of the rate-limiting step. The first is the concerted process. Since there is only one step, and it involves a molecule of both the substrate and the nucleophile, this step is bimolecular.

> S$_N$2 mechanism
>
> RX + Nu$^-$ $\longrightarrow$ RNu + X$^-$
>
> Concerted (one step) bimolecular
>
> $J = k[\text{RX}][\text{Nu}^-]$

Thus this mechanism is known as S$_N$2 (Substitution, Nucleophilic, Bimolecular). This is the more common mechanism.

The other less common mechanism is a two-step dissociative mechanism with the first step, which doesn't involve the nucleophile, being rate-limiting.

> S_N1 mechanism
>
> $RX \longrightarrow R^+ + X^-$ Slow
>
> $R^+ + Nu^- \longrightarrow RNu$ Fast
>
> Two steps, rate-limiting step unimolecular
>
> $J = k[RX]$

Since the first, rate-limiting step, only involves the substrate, it is a unimolecular process and is thus given the name S_N1 (Substitution, Nucleophilic, Unimolecular).

🔵 Is it reasonable that the first step is the slower of the two?

🔵 As the first step involves the formation of charged species by breaking a covalent bond and the second step involves the reaction of oppositely charged ions, it does seem a reasonable proposition, that the first, bond-breaking, step will be the slower.

Effectively, as soon as the carbocation is formed in the slow first step, it reacts very quickly with the nucleophile in the fast second step. So the overall rate of reaction is going to be influenced only by the rate of the first step.

If you look again at the qualitative rate observations in Table 4.1, you should find them a little less puzzling. The problems arose because we were considering that the reactions proceeded by the same mechanism. In the S_N2 mechanism, the nucleophilicity of the reagent is important in determining the rate of the reaction. Table 4.2 repeats the information in Table 4.1 with an additional column in which the reaction mechanisms are classified.

Table 4.2 Rate data and classification for reactions $RBr \longrightarrow ROCH_2CH_3$

Substrate, RBr	Conditions	Rate	Type of reaction
CH_3CH_2Br	CH_3CH_2OH/boil	50% complete in 4 days	S_N2
CH_3CH_2Br	$NaOCH_2CH_3$/solvent, CH_3CH_2OH	complete in a few minutes	S_N2
$(CH_3)_3CBr$	CH_3CH_2OH/boil	complete in a few minutes	S_N1

🔵 Is the nucleophilicity of the reagent important in determining the rate of a reaction proceeding by an S_N1 mechanism?

🔵 No, because the nucleophile is involved only in the second step of the reaction, which is comparatively fast, irrespective of the nucleophilicity.

When reactions proceed by the S_N1 mechanism, the nucleophilicity of the reagent is not important. Carbocations such as Structure **4.2** are such reactive electrophiles that they react rapidly with any nucleophile.

Bromoethane reacts with ethanol (or with sodium ethoxide) by the S_N2 mechanism. Because ethanol is a poor nucleophile this reaction is very slow. On the other hand,

2-bromo-2-methylpropane reacts by the S_N1 mechanism; as soon as a carbocation is formed it reacts rapidly, even with a poor nucleophile such as ethanol. The two processes are shown in Scheme 4.1. Note that with ethanol as the nucleophile, the first-formed product of substitution is protonated. However, proton transfer reactions are generally very fast, and loss of H^+ gives the product.

S_N2:

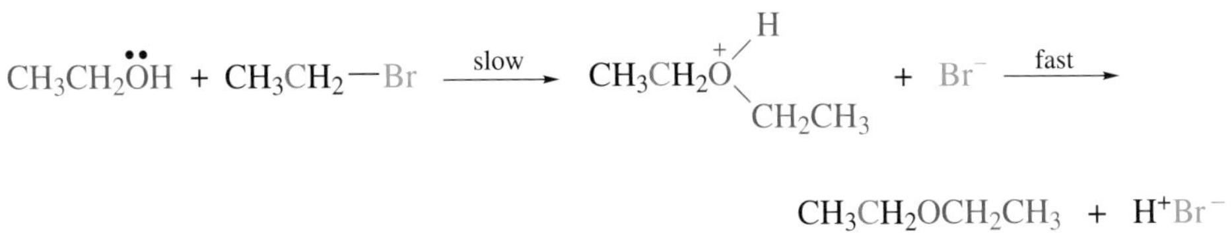

S_N1:

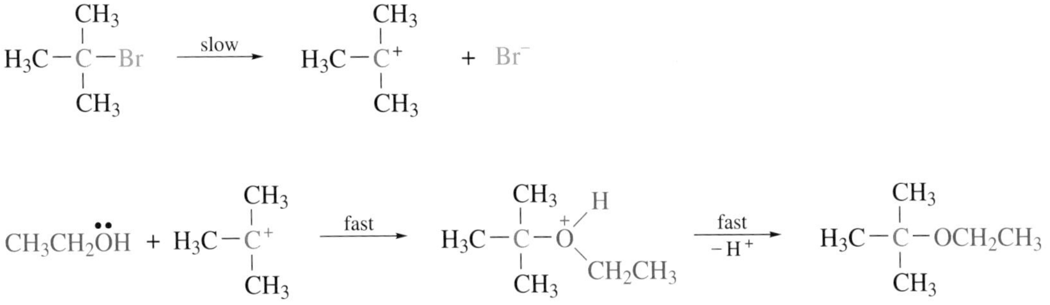

(Scheme 4.1)

You may have wondered why a stable molecule, such as $(CH_3)_3CBr$, should suddenly fall apart to give a carbocation. In fact, such processes do occur, albeit fairly rarely. However, the reaction is reversible

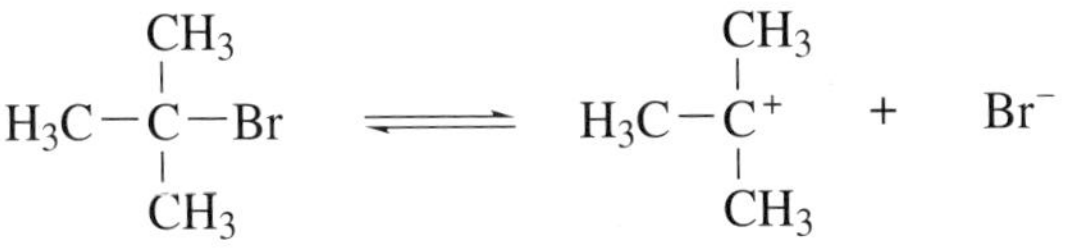

so in the absence of nucleophiles, the oppositely charged ions combine to reform the bromoalkane. (This is like the situation with water, in which only a very small proportion of the water is in the ionic state, as H^+ and ^-OH, but there are *some* ions present all the time.) When a nucleophile is present the carbocation can be intercepted to give the product. But this hasn't really answered the question of how ionization happens. If a bromoalkane molecule collides with another molecule, such as a solvent molecule, sufficient energy may be transferred to the substrate for it to overcome the energy barrier to this ionization. The overall process is slow because this energy barrier is high, and only a few of the molecules at any one time have sufficient energy to surmount it. The actual numbers of these molecules that gain sufficient energy to overcome the barrier at any one time will depend on, amongst other things, the number of reactant molecules present, and this is why the rate of reaction depends on the concentration of the substrate.

We still haven't explained why different mechanisms operate with different substrates, nor have we explained the stereochemical observations discussed in Section 2.1, Example 2.3. But the basis for these explanations has been laid, and we will reveal all in the CD-ROM.

COMPUTER ACTIVITY 4.1 Kinetics and stereochemistry

At some point in the near future you should study *Kinetics and stereochemistry* in the CD-ROM that accompanies this book. This reviews the material on the kinetics and mechanism of S_N reactions and describes the key features of S_N1 and S_N2 reactions including the stereochemistry. If you cannot study the CD-ROM immediately, don't worry, you can read on, since the text in the next few sections does not rely on the new concepts taught on the CD-ROM.

4.3 Summary of Section 4

1 Kinetic studies of S_N reactions support the idea put forward by Hughes and Ingold that there are two distinct mechanisms for S_N reactions. In the first mechanism, the bonds are made and broken in a concerted manner, so that reactants are converted into products in one step. For example

$$RX \quad + \quad Nu^- \quad \longrightarrow \quad RNu \quad + \quad X^-$$

via a transition state $[Nu \cdots R \cdots X]^-$

This is a bimolecular mechanism, and as expected, kinetic studies show that the reaction is second-order, with a rate given by $k[RX][Nu^-]$. This mechanism is known as S_N2 (substitution, nucleophilic, bimolecular), and it is the most commonly encountered substitution mechanism.

2 The less common S_N mechanism is a two-step process proceeding by way of a carbocation reaction intermediate. For example

$$R{-}X \quad \xrightarrow{\text{slow}} \quad R^+ \quad + \quad X^-$$

$$Nu^- \quad + \quad R^+ \quad \xrightarrow{\text{fast}} \quad R{-}Nu$$

The first step is slow and rate-limiting. Only the substrate is involved, so this process is unimolecular and shows first-order kinetics, with a rate given by $k[RX]$.

This mechanism is known as S_N1 (substitution, nucleophilic, unimolecular). Because carbocations are such reactive electrophiles, even poor nucleophiles such as alcohols and water, react with them.

QUESTION 4.1

Draw out the mechanism of the following reactions

(i) $CH_3{-}CH_2{-}Br$ plus HO^- via an S_N2 reaction

(ii) $H_3C{-}\overset{\displaystyle CH_3}{\underset{\displaystyle CH_3}{\overset{|}{\underset{|}{C}}}}{-}Cl$ plus NH_3 via an S_N1 reaction

5

S$_N$2 VERSUS S$_N$1

You have seen that there are two distinct mechanisms by which the nucleophilic substitution reaction can proceed — the S$_N$2 mechanism, and the less common S$_N$1 mechanism. One question that we have not yet answered is 'Is it possible to make one mechanism more likely to occur than the other and, if so, how?' This question would be crucial if you wanted to synthesize a molecule with a particular stereochemistry.

The ideal situation would be one in which any substrate could be made to react with any nucleophile by either mechanism, simply by altering the reaction conditions. As you will see later, modification of the reaction conditions can sometimes lead to a change in the mechanism, but the major factor determining which mechanism operates is the structure of the substrate.

5.1 The effect of substrate structure

There are two distinct mechanisms of nucleophilic substitution, and so far we have considered only examples in which one mechanism operates to the exclusion of the other.

Although we have studied S$_N$1 and S$_N$2 reactions in isolation, in practice, they usually occur in competition.

> For any substitution reaction, the overall rate of reaction is the sum of the rates of the S$_N$1 and S$_N$2 processes
>
> $$J_{overall} = (\text{rate of } S_N1) + (\text{rate of } S_N2)$$
> $$= k_{S_N1}[RX] + k_{S_N2}[RX][Nu^-]$$
> $$= \left(k_{S_N1} + k_{S_N2}[Nu^-]\right)[RX]$$
>
> S$_N$2 predominates when $k_{S_N1} \ll k_{S_N2}[Nu^-]$.
>
> S$_N$1 predominates when $k_{S_N1} \gg k_{S_N2}[Nu^-]$.
>
> When k_{S_N1} and $k_{S_N2}[Nu^-]$ have similar values, substitution will occur by a mixture of S$_N$1 and S$_N$2 mechanisms.

So, in order to make predictions about which mechanism will predominate for a given substrate, we need to have a feeling for the way in which the rate constants, k_{S_N1} and k_{S_N2} vary with the substrate structure. This information can be obtained by carrying out a series of reactions in which the only variable is the structure of the substrate; that is, the nucleophile, leaving group, concentrations, solvent and temperature are all kept the same. Under these conditions, differences in rates of reaction are due to differences in rate constants for the particular mechanism in operation. Experimental data for a series of reactions proceeding by the S$_N$2 mechanism are given in Table 5.1.

Table 5.1 Relative rates of S$_N$2 reactions of various bromoalkanes with hydroxide ion

Compound	Relative rate
CH_3Br	30
CH_3CH_2Br	1.0
$(CH_3)_2CHBr$	0.02
$(CH_3)_3CBr$	does not react by S$_N$2

As you can see, increasing the number of alkyl groups attached to the electrophilic carbon atom slows down the reaction. As a general rule, for an S_N2 mechanism, CH_3X reacts faster than primary substrates, which react faster than secondary substrates; tertiary substrates rarely undergo S_N2 reactions:

rate of S_N2:

$$CH_3X > RCH_2X > R_2CHX >> R_3CX$$

The variation of the rate of the S_N1 process with substrate structure has also been studied and some typical results are shown in Table 5.2. There is no correlation between the relative rates in this table and those in Table 5.1.

With S_N1 processes the opposite trend is observed, increasing the number of alkyl groups attached to the electrophilic carbon atom increases the rate of reaction. As a general rule, for an S_N1 mechanism, tertiary substrates react faster than secondary substrates, which react faster than primary substrates; CH_3X rarely undergoes S_N1 reactions:

rate of S_N1:

$$R_3CX > R_2CHX > RCH_2X > CH_3X$$

Table 5.2 Relative rates for the reaction of bromoalkanes with water by the S_N1 mechanism

Compound	Relative rate
CH_3Br	1.0
CH_3CH_2Br	1.7
$(CH_3)_2CHBr$	45
$(CH_3)_3CBr$	10^6

The reasons for these trends are explained on the CD-ROM. However, for the moment we will concentrate on predicting which mechanism will be preferred, based on substrate structure. If you look at the relative rates of reactions proceeding by S_N1 and S_N2 mechanisms, you will see that tertiary substrates react fastest by the S_N1 mechanism whereas primary substrates react fastest by the S_N2 mechanism:

rate of S_N2 increases

←————————————————————

CH_3X RCH_2X R_2CHX R_3CX

————————————————————→

rate of S_N1 increases

In the kinetic studies described in Tables 5.1 and 5.2, the reactions were forced to follow either the S_N1 or the S_N2 mechanism to yield the relevant rate data. In general, as we said earlier, the two mechanisms compete. This means that the faster mechanism predominates. So, methyl and primary alkyl substrates generally react by the S_N2 mechanism, and tertiary substrates almost always react by the S_N1 mechanism. Secondary substrates can react via either mechanism depending upon the conditions. Exceptions to this rule include compounds with one or more phenyl groups directly attached to the carbon atom undergoing substitution.

As a rule of thumb, one phenyl group has a similar effect to two alkyl groups.

All this information about the mechanisms that normally operate for a given substrate is summarized in Table 5.3.

Table 5.3 Normal substitution mechanism for different substrates (R = alkyl)

S$_N$2 preferred	S$_N$2 or S$_N$1	S$_N$1 preferred
CH$_3$X, RCH$_2$X	R$_2$CHX	R$_3$CX
	C$_6$H$_5$CH$_2$X	(C$_6$H$_5$)$_3$CX, (C$_6$H$_5$)$_2$CRX
		(C$_6$H$_5$)$_2$CHX, C$_6$H$_5$CHRX

The synthetic chemist has very little control over the reaction mechanism when the substrates are of the type shown in the S$_N$2-preferred and S$_N$1-preferred columns of Table 5.3. The major area where control *can* be exercised is with secondary substrates, and these often undergo substitution by a mixture of both mechanisms, as the rates for each are similar. However, if it is important to ensure that one mechanism predominates, this can usually be done by a careful choice of conditions, such as the concentration of the nucleophile.

5.2 The effect of the nucleophile

Returning to the equation for the overall rate of a substitution reaction, you can see that one important factor is the concentration of the nucleophile

$$J_{overall} = (\text{rate of S}_N1) + (\text{rate of S}_N2)$$
$$= k_{S_N1}[RX] + k_{S_N2}[RX][Nu^-] \tag{5.1}$$

Equation 5.1 shows that if the rates of S$_N$1 and S$_N$2 reactions are broadly similar, the rate of the S$_N$2 process can be increased relative to the S$_N$1 process by increasing the concentration of the nucleophile. With a low concentration of nucleophile the reaction may occur by both S$_N$1 and S$_N$2 mechanisms, but with a high concentration of nucleophile, $k_{S_N2}[Nu^-]$ can be much greater than k_{S_N1} and so reaction by the S$_N$2 mechanism predominates.

The nucleophilicity of the nucleophile can also have an effect on the rate of an S$_N$ reaction.

● Would you expect an increase in the nucleophilicity of the nucleophile to increase the rate of a reaction occurring by the S$_N$1 mechanism?

● No. The rate of an S$_N$1 reaction depends only on the rate of ionization of the substrate. The second step will be fast for all nucleophiles.

However, the rates of reactions proceeding by the S$_N$2 mechanism *are* dependent on the nucleophilicity of the nucleophile, with the reactivity of the nucleophile being reflected in the value of the rate constant, k_{S_N2} for a given reaction.

So, for a given substrate capable of reacting by either mechanism, the type of nucleophile can sometimes be altered to favour one mechanism over the other. You can see this with Reactions 5.2 and 5.3:

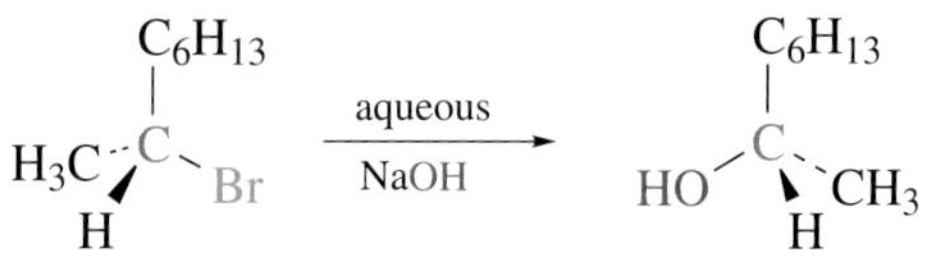

$$\underset{\substack{H_3C \diagdown C \diagup Br \\ H}}{\overset{C_6H_{13}}{|}} \quad \xrightarrow[\text{NaOH}]{\text{aqueous}} \quad \underset{\substack{HO \diagup C \diagdown CH_3 \\ H}}{\overset{C_6H_{13}}{|}} \qquad (5.2) \quad \square$$

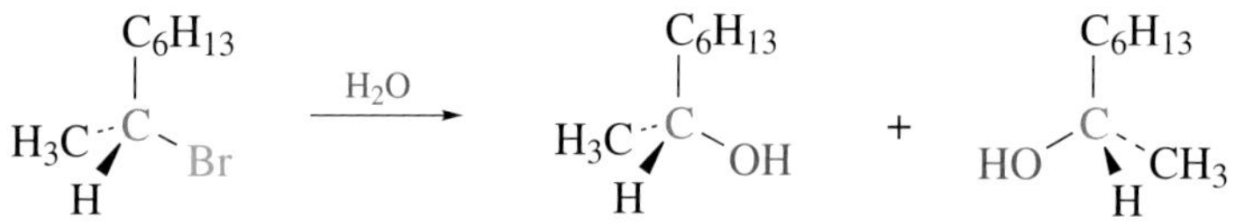

$$\underset{\substack{H_3C \diagdown C \diagup Br \\ H}}{\overset{C_6H_{13}}{|}} \quad \xrightarrow{H_2O} \quad \underset{\substack{H_3C \diagdown C \diagup OH \\ H}}{\overset{C_6H_{13}}{|}} \quad + \quad \underset{\substack{HO \diagup C \diagdown CH_3 \\ H}}{\overset{C_6H_{13}}{|}} \qquad (5.3) \quad \square$$

● What are the nucleophiles in Reactions 5.2 and 5.3?

● Hydroxide ($^-$OH) in Reaction 5.2, and water in Reaction 5.3.

Hydroxide has a high nucleophilicity (k_{S_N2} is large), so Reaction 5.2 proceeds by the S_N2 mechanism only. Water has a low nucleophilicity (k_{S_N2} is small), so in Reaction 5.3 the rates of the S_N1 and S_N2 processes are similar, and the reaction proceeds by a mixed mechanism. So, a high concentration of a reactive nucleophile favours the S_N2 mechanism.

COMPUTER ACTIVITY 5.1 Inductive and resonance effects

At some point in the near future you should study *Inductive and resonance effects* on the CD-ROM that accompanies this book. This reviews the material covered on S_N1 versus S_N2, and explains the origin of the trends observed.

5.3 Summary of Section 5

1 The mechanisms usually observed for the various types of substrate are given in Table 5.3.

2 Primary substrates usually react by the S_N2 mechanism. Tertiary substrates usually react by the S_N1 mechanism. Secondary substrates can react by either mechanism.

3 With secondary substrates the S_N2 mechanism can usually be favoured by using a high concentration of a reactive nucleophile.

QUESTION 5.1

In each of the following reactions predict whether an S_N1 or S_N2 mechanism (or both) will be followed. Give your reasoning.

(i) $C_6H_5CH_2CH_2CH_2Cl \quad + \quad Na^+I^-$

(ii) $CH_3CH_2CH_2Br \quad + \quad Na^+CN^-$

(iii) $\underset{\substack{| \\ Br}}{\overset{\substack{CH_3 \\ |}}{C_6H_5-C-CH_3}} \quad + \quad Ag^{+-}O-\overset{\overset{\textstyle O}{\|}}{C}-CH_3$

(iv) $\underset{}{\overset{\substack{I \\ |}}{CH_3-CH-CH_3}} \quad + \quad Na^+CN^- \text{(in high concentration)} \qquad \square$

CONCLUDING REMARKS

6

At the start of this part of this book we discussed the importance of organic reactions, and we spoke of the problems associated with planning and predicting the outcome of such reactions. We posed a number of problems, and stressed the need for an understanding of the mechanisms of organic reactions. If you go back and look at Section 2.1 you should now understand most of the observations made there. Example 2.2 should still be puzzling, but we shall deal with elimination reactions in Part 3. However, once you have finished the CD-ROM you should be able to understand all the other observations, which must have seemed perplexing at the time.

This part of this book, including the CD-ROM Activities, is long but it's important. Not only have we introduced substitution reactions, but we have also introduced mechanistic organic chemistry. You will meet curly arrows, nucleophiles, electrophiles and leaving groups again and again and again in organic chemistry! An understanding of reaction mechanisms is invaluable because it enables organic chemists to make predictions about reactions that have not yet been carried out. When we discuss synthesis later, you will see how important it is to be able to predict reaction mechanisms accurately!

QUESTION 6.1

Complete the following definitions by choosing one of the alternative endings.

(i) Heterolysis of a covalent bond produces...

    (a) radicals;

    (b) ions.

(ii) The S_N2 mechanism involves...

    (a) one step;

    (b) two steps.

(iii) The reactivity and concentration of a nucleophile affect the rate of...

    (a) a reaction proceeding by the S_N2 mechanism;

    (b) a reaction proceeding by the S_N1 mechanism.

LEARNING OUTCOMES FOR PART 2

Now that you have completed Part 2 *The Mechanism of Substitution*, you should be able to do the following:

1   Recognize valid definitions of and use in a correct context the terms, concepts and principles in the following table (All questions).

List of scientific terms, concepts and principles used in Part 2 of this book.

Term	Page number	Term	Page number
addition reaction	137	nucleophilicity	163
bond heterolysis	145	nucleophilic substitution reaction	153
bond homolysis	145	polarity of a bond	CD
carbocation	171	radical	145
chain reaction	146	radical reaction	145
curly arrow	CD	reaction mechanism	144
electron-withdrawing effect	CD	reagent	136
electrophile	152	resonance effect	CD
electrophilic atom	152	resonance form	CD
elimination reaction	137	retention of configuration	CD
inductive effect	CD	S_N reaction	153
inversion of configuration	CD	spectator ion	151
ionic reaction	145	steric hindrance	CD
leaving group	152	substitution reaction	137
nucleophile	151	substrate	136

2   Given the reactants and products of an organic reaction,

   (a)   classify the reaction as a substitution, elimination, or addition reaction;

   (b)   decide whether an ionic or a radical mechanism is operating.
         (Questions 1.1, 2.1 and CD-ROM Questions)

3   Given the reactants and products of a nucleophilic substitution reaction,

   (a)   specify the nucleophile, leaving group, and electrophilic carbon centre;

   (b)   if possible, decide whether an S_N1 or S_N2 mechanism is operating.
         (Questions 3.1, 5.1 and CD-ROM Questions)

4   Use curly arrows to illustrate reaction mechanisms. (CD-ROM Questions)

5   Write balanced equations to illustrate S_N reactions. (Questions 3.2, 4.1 and CD-ROM Questions)

6   Predict the reaction mechanism, and hence the structure of the products, given the reactants in a nucleophilic substitution reaction. (Questions 3.2, 5.1 and CD-ROM Questions)

7   Decide whether a proposed S_N reaction will either go to completion, or give a mixture of reactants and products, or not go at all. (Question 3.3 and CD-ROM Questions)

8   Comment on the relative rates of two S_N reactions, basing your arguments on the relative reactivities of the nucleophiles. (Question 3.4 and CD-ROM Questions)

9   Suggest the best reagent and reaction conditions for carrying out a given transformation. (Question 5.1 and CD-ROM Questions)

QUESTIONS: ANSWERS AND COMMENTS

QUESTION 1.1 (*Learning Outcome 2*)

In each case, we focus on what is happening in the organic species.

(i)   Substitution (Cl for H)

(ii)   Substitution (CN for Br)

(iii)   Addition (of H and OH)

(iv)   Addition (of H and H)

(v)   Elimination (of Cl and Cl)

(vi)   Substitution (Cl for OH)

(vii)   Substitution (NO_2 for H)

(viii) Addition (of H and CN)

(ix)   Substitution (Br for OH)

QUESTION 2.1 (*Learning Outcome 2*)

(i)   Ionic substitution; an ionic reagent ($Na^+\ ^-OH$) is employed, and substitution occurs specifically of the bromine functional group.

(ii)   Radical substitution; a covalent reagent (Cl_2) is employed, a hydrogen atom is substituted, light is necessary, and a mixture of products is obtained.

(iii)   Ionic addition; reaction occurs at the C=O functional group and an ionic reagent is employed ($Na^+\ ^-BH_4$).

(iv)   Ionic addition; an ionic reagent is used, and reaction occurs only at the C≡C functional group.

(v)   Ionic elimination; reaction occurs at the site of the OH functional group (with an elimination reaction, the adjacent site is also involved).

QUESTION 3.1 (*Learning Outcome 3*)

The nucleophiles, leaving groups and spectator ions are given in Table Q.1 below. The electrophilic carbon atom is *always* the one bonded to the leaving group.

Table Q.1   Answer to Question 3.1

	Nucleophile	Leaving group	Spectator ion
(i)	$^-SCH_3$	Cl	Na^+
(ii)	^-CN	Cl	K^+
(iii)	I^-	Br	Na^+

QUESTION 3.2 (*Learning Outcomes 5 and 6*)

(i) (cyclohexyl)—CH_2OSO_2—(benzene ring)—CH_3 + $Li^+ Br^-$ $\longrightarrow$

(cyclohexyl)—CH_2Br + $Li^+ {}^-OSO_2$—(benzene ring)—CH_3

(ii) $C_6H_5CH_2Br$ + $Na^+ {}^-SH$ $\longrightarrow$ $C_6H_5CH_2SH$ + $Na^+ Br^-$

(iii) $\overset{..}{N}H_3$ + CH_3I $\longrightarrow$ $CH_3\!-\!\overset{+}{N}H_3\ I^-$

(iv) CH_3I + $CH_3C\equiv C^- Na^+$ $\longrightarrow$ $CH_3C\equiv CCH_3$ + $Na^+ I^-$

Notice that in each transformation the same atoms and overall charges are present on either side of the transformation arrow.

QUESTION 3.3 (*Learning Outcome 7*)

(i) This reaction won't go, because CN is not a good leaving group.

(ii) This reaction should go to completion, because there is a good nucleophile ($^-SCH_3$) and a good leaving group.

(iii) This reaction won't go, as OH is a poor leaving group.

(iv) This reaction will go, as I^- is a good nucleophile and Cl is a good leaving group. But because Cl^- is also a good nucleophile you should have predicted that a mixture of reactants and products will be obtained.

(v) This reaction will go, because the nucleophile is Cl^- and the leaving group is OH_2. Remember that the hydroxyl group acts as a leaving group only in the presence of acid.

QUESTION 3.4 (*Learning Outcome 8*)

(i) Reaction (a) will proceed faster, because the iodide anion is much more reactive as a nucleophile than the chloride anion (nucleophilicity increases with atomic number in a Group of the Periodic Table).

(ii) Reaction (a) will proceed faster. Anions are better nucleophiles than the corresponding protonated compounds; and in this case, ^-OH is a better nucleophile than H_2O.

(iii) From your present knowledge you cannot say, because the nucleophile is the same in both reactions but the structure of the substrate has changed. In fact, reaction (b) will be much slower than (a), as you will see in Section 5.1.

QUESTION 4.1 (*Learning Outcome 5*)

(i)   This is an S_N2 reaction so it occurs in one step

$$CH_3CH_2Br \quad + \quad HO^- \quad \longrightarrow \quad CH_3CH_2OH \quad + \quad Br^-$$

via an activated complex

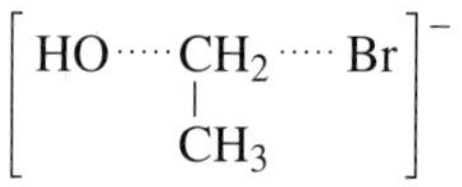

$$\left[\begin{array}{c} HO \cdots CH_2 \cdots Br \\ | \\ CH_3 \end{array} \right]^-$$

(ii) This is an S_N1 reaction so it occurs in two steps

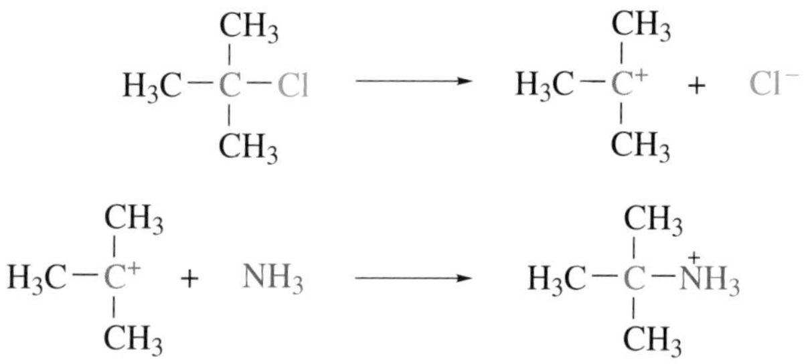

$$\begin{array}{c} CH_3 \\ | \\ H_3C-C-Cl \\ | \\ CH_3 \end{array} \longrightarrow \begin{array}{c} CH_3 \\ | \\ H_3C-C^+ \\ | \\ CH_3 \end{array} + \quad Cl^-$$

$$\begin{array}{c} CH_3 \\ | \\ H_3C-C^+ \\ | \\ CH_3 \end{array} + \quad NH_3 \quad \longrightarrow \begin{array}{c} CH_3 \\ | \\ H_3C-C-\overset{+}{N}H_3 \\ | \\ CH_3 \end{array}$$

QUESTION 5.1 (*Learning Outcomes 3, 6 and 9*)

(i)   The reaction of a primary substrate with a good nucleophile, such as iodide, would be expected to go by an S_N2 mechanism. However, chloride is also a good nucleophile, and an equilibrium would be expected.

(ii)   Once again, the S_N2 mechanism is expected with a primary substrate and a good nucleophile.

(iii) This reaction would almost certainly proceed exclusively by an S_N1 mechanism. All the conditions (a tertiary substrate, a fairly poor nucleophile) favour this mechanism.

(iv)   Secondary substrates can react by either the S_N1 or the S_N2 mechanism. However, the use of a good nucleophile, in high concentration should favour the S_N2 mechanism.

QUESTION 6.1 (*Learning Outcome 1*)

(i) b; (ii) a; (iii) a.

Part 3

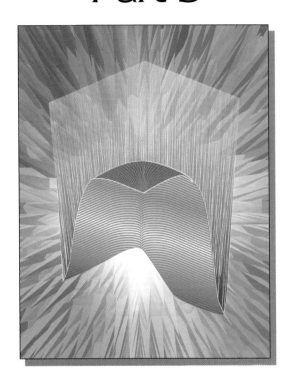

Elimination: Pathways and Products

edited by Peter Taylor
from work authored by Richard Taylor

INTRODUCTION: β-ELIMINATION REACTIONS

Many very important compounds (for example ethene, Figure 1.1) contain carbon–carbon double bonds.

To make such compounds we need to be able to introduce these double bonds into molecules in a controlled manner. One way of doing this is by elimination. Cast your mind back to the classification of organic reactions that we set out at the start of Part 2 of this book.

● What is the definition of an elimination reaction?

● An **elimination reaction** involves the removal of two atoms or groups from a molecule.

There are many types of elimination reaction, the most common being that in which the two atoms or groups are eliminated from adjacent carbon atoms. Conventionally, these two carbon atoms are labelled α and β, and the reaction is called **β-elimination**

Effectively two singly bonded atoms or groups are removed from adjacent carbon atoms, and a double bond is formed between the carbon atoms to which the eliminated groups had been attached. Some examples of β-eliminations are listed in Table 1.1.

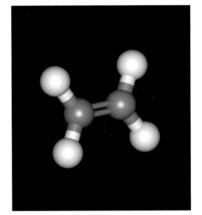

Figure 1.1
Ball and stick model of ethene: notice the planar nature of this molecule.

Table 1.1 Some typical elimination reactions

(a) $CH_3CH_2CH_2Br + Na^{+-}OH \longrightarrow CH_3CH=CH_2 + Na^+Br^- + H_2O$

(b) [cyclohexyl OSO₂C₆H₄CH₃] $+ Na^{+-}OCH_2CH_3 \longrightarrow$ [cyclohexene] $+ CH_3CH_2OH + Na^{+-}OSO_2C_6H_4CH_3$

(c) $C_6H_5CH_2CH_2\overset{+}{N}(CH_3)_3\ {}^-OH \longrightarrow C_6H_5CH=CH_2 + (CH_3)_3N + H_2O$

(d) $CH_3C{=}CH_2$ (Br) $+ Na^{+-}NH_2 \longrightarrow CH_3C{\equiv}CH + Na^+Br^- + NH_3$

(e) $CH_3CH{-}CHCH_3$ (Br Br) $+ Zn \longrightarrow CH_3CH=CHCH_3 + ZnBr_2$

(f) $BrH_2C{-}\overset{CH_3}{\underset{CH_3}{C}}{-}COOH + Na^{+-}OH \longrightarrow H_2C=C\overset{CH_3}{\underset{CH_3}{}} + Na^+Br^- + CO_2 + H_2O$

What atoms or groups are eliminated in reactions (a), (b) and (c) in Table 1.1?

(a) H and Br,

(b) H and $CH_3C_6H_4SO_3$,

(c) H and $N(CH_3)_3$.

In all three reactions one of the eliminated groups is a hydrogen atom. This is the most common type of b-elimination

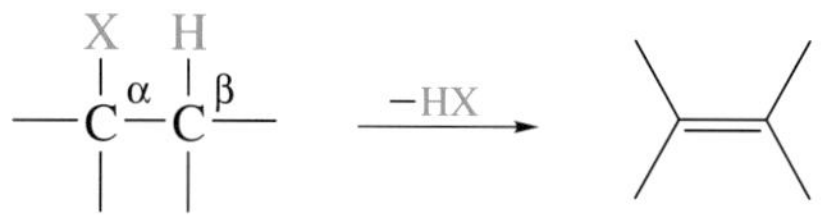

We shall discuss this type of β-elimination first, before going on to consider the other types of elimination reaction exemplified by reactions (d), (e) and (f) in Table 1.1.

BOX 1.1 The industrial importance of alkenes

Plastics are one of the most important products of the chemical industry and the majority of these are prepared from alkenes. Table 1.2 lists some of the typical alkene monomers.

Alkenes are important intermediates in the organic chemical industry since many products can be made from them. For example, 66 million tonnes of ethene are manufactured each year worldwide. Notice that an alternative name for an alkenyl compound is a vinyl compound.

Table 1.2 Typical alkene monomers together with the plastics formed from them

Name	Structure	Common name	Plastic product	
ethene	$H_2C{=}CH_2$	ethylene	polythene	💻
chloroethene	$H_2C{=}CHCl$	vinyl chloride	PVC	💻
phenylethene	$C_6H_5CH{=}CH_2$	styrene	polystyrene	💻
tetrafluoroethene	$F_2C{=}CF_2$	-	PTFE (Teflon®)	💻
ethenyl ethanoate	$H_2C{=}CH{-}OCOCH_3$	vinyl acetate	PVA	💻
methyl 2-methylpropenoate	$H_2C{=}C\begin{smallmatrix}CH_3\\COOCH_3\end{smallmatrix}$	methyl methacrylate	poly(methyl methacrylate) (perspex) 💻	

1.1 The mechanisms of β-elimination reactions

Let's consider the possible mechanisms of β-elimination by looking at a typical reaction, the formation of an alkene by treating a haloalkane with sodium hydroxide

$$CH_3CH_2CH_2Br \ + \ Na^+{}^-OH \ \longrightarrow \ CH_3CH{=}CH_2 \ + \ Na^+Br^- \ + \ H_2O$$

⬤ What role do you think the bromine plays in this reaction?

⬤ It acts as a leaving group.

All the elimination reactions discussed in this book proceed by ionic mechanisms, and the simplest such mechanism is a concerted one-step process reminiscent of the S_N2 mechanism (Reaction 1.1)

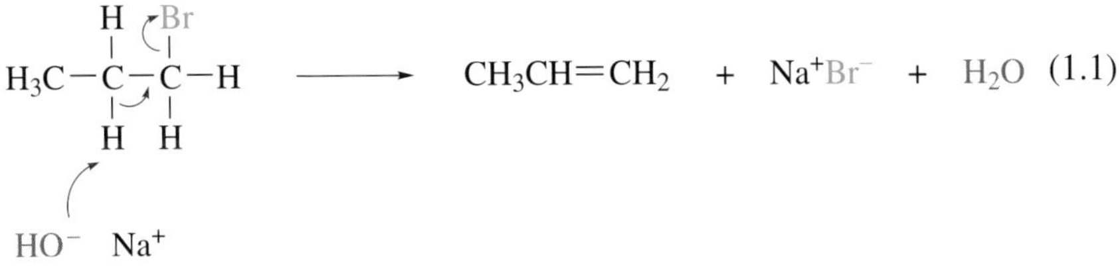

⬤ What role is the hydroxide anion playing in Reaction 1.1?

⬤ It is forming a new bond to a hydrogen atom. In other words, it is acting as a *base*.

A **base** is defined as a species that accepts protons. In the above reaction the ^-OH acts as a base and accepts a proton from the bromoalkane. So, in elimination reactions the ^-OH acts as a base, whereas, as you saw in Part 2, in substitution reactions the same ion acts as a nucleophile (attacking and forming a bond to a carbon atom). Don't worry too much about the interplay between elimination and substitution reactions just now; we shall discuss this topic in more detail in Section 4. Just remember that ^-OH can act in two different ways, depending on the type of reaction. In the mechanism shown in Reaction 1.1, the base removes the proton; the two electrons from the carbon–hydrogen bond form the new bond of the double bond; and the carbon–bromine bond breaks heterolytically, the two electrons going to the bromine to give a bromide anion.

⬤ What is the molecularity of this one-step reaction?

⬤ Two molecules (base and the bromoalkane) are involved in the only step of the reaction, and so it is *bimolecular*.

This mechanism is known as the **E2 mechanism**, with 'E' standing for elimination and '2' for bimolecular

Generalised E2 mechanism (B = a base)

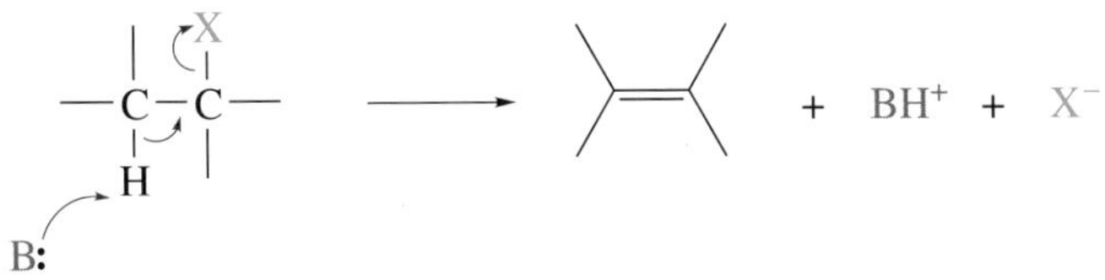

⬤ By analogy with the S_N1 mechanism, can you think of a two-step elimination mechanism?

⬤ One possibility involves the ionization of the leaving group in the first, rate-limiting step of the reaction, and then removal of the proton and formation of the double bond in a second, fast, step

Generalised E1 mechanism (B = a base)

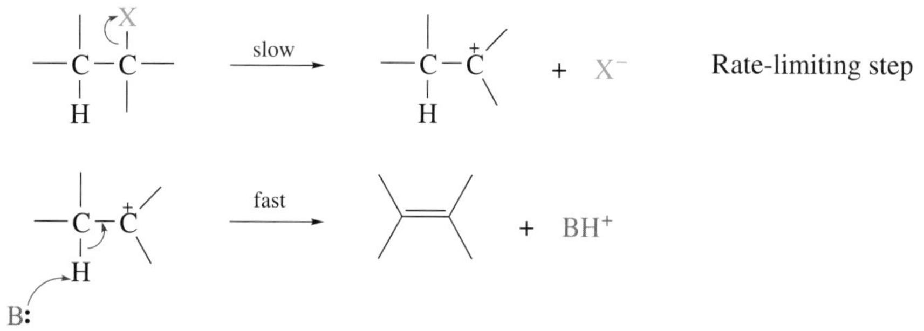

An alternative two-step mechanism that might have occurred to you involves the removal of the proton in the first step, followed by the expulsion of the leaving group. This mechanism is sometimes encountered, but not often enough for us to worry about it here.

The molecularity of the rate-limiting step is 1 (only the substrate is involved) and so this process is called the **E1 mechanism**. Notice that the carbocation could also undergo a substitution reaction (that is, an S_N1 mechanism could operate). We shall discuss this further in Section 4.

⬤ How would you determine which elimination mechanism was operating in a given reaction?

⬤ The first thing to do would be to carry out a kinetic investigation as described in Part 1 of this book.

Reactions proceeding by the E1 mechanism would show first-order kinetics, and reactions proceeding by the E2 mechanism would show second-order kinetics

El mechanism	$J = k[\text{substrate}]$
E2 mechanism	$J = k'[\text{substrate}][\text{base}]$

Investigations such as this reveal that the E2 mechanism is by far the more common mechanism of elimination.

1.2 Summary of Section 1

1 Elimination reactions involve the removal of two atoms or groups from a molecule.

2 When the eliminated groups are on adjacent carbon atoms, the degree of unsaturation is increased; for example

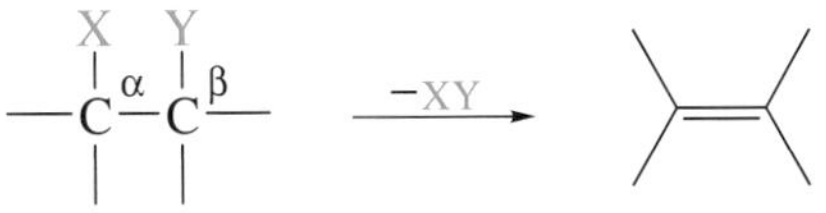

This type of elimination is known as β-elimination.

3 β-eliminations often involve the elimination of H and X from a molecule, where X is a good leaving group. The reaction is brought about by treating the substrate with a base

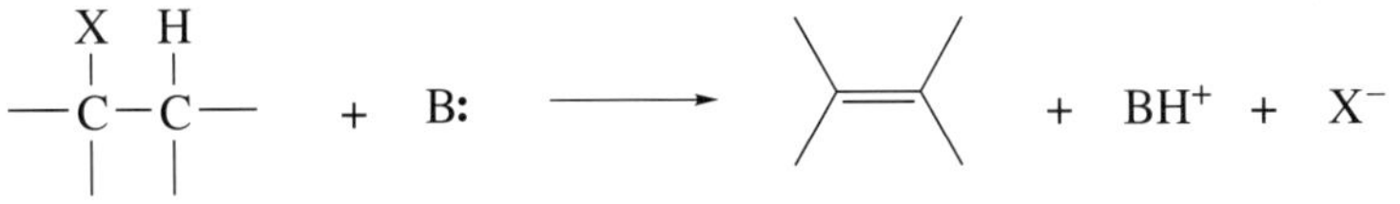

4 β-eliminations usually proceed by either the E2 or the E1 mechanism. The more common E2 mechanism is a concerted, one-step process. The E1 mechanism involves two steps, a carbocation intermediate being formed in the slow (rate-limiting) first step.

QUESTION 1.1

Use curly arrows to illustrate the mechanisms of the following reactions.

(a) $ClCH_2CH_2Cl$ + $Na^+ \ ^-OH$ $\longrightarrow$ $ClCH{=}CH_2$ + Na^+Cl^- + H_2O

by an E2 mechanism

(b) $OSO_2C_6H_4CH_3$

+ $Na^+ \ ^-OCH_3$ $\longrightarrow$ + CH_3OH + $Na^+ \ ^-OSO_2C_6H_4CH_3$

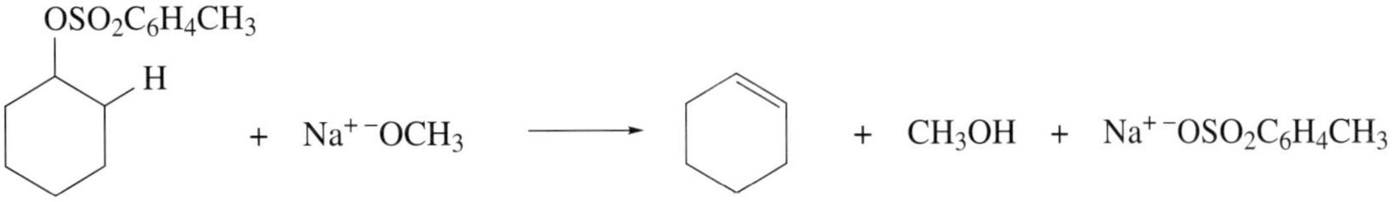

by an E2 mechanism

(c) $(CH_3)_3CBr$ + $Na^+ \ ^-OCH_2CH_3$ $\longrightarrow$ $(CH_3)_2C{=}CH_2$ + Na^+Br^- + CH_3CH_2OH

by an E1 mechanism

THE E2 MECHANISM

2

2.1 The scope of the E2 mechanism

We have said that the most commonly encountered mechanism of elimination is the concerted, one-step, E2 mechanism

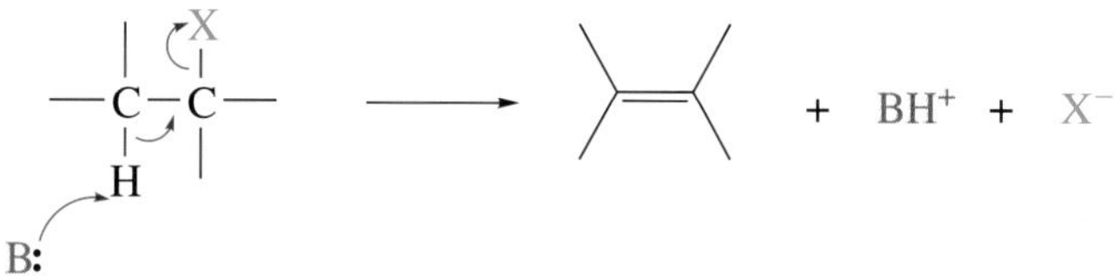

Table 2.1 Leaving groups (X) in E2 reactions

Substrate (RX)	X⁻ in products
R—Cl	Cl⁻
R—Br	Br⁻
R—I	I⁻
R—OSO$_2$C$_6$H$_4$CH$_3$	CH$_3$C$_6$H$_4$SO$_3$⁻
R—$\overset{+}{\text{N}}$(CH$_3$)$_3$	(CH$_3$)$_3$N
R—$\overset{+}{\text{S}}$(CH$_3$)$_2$	(CH$_3$)$_2$S

The leaving groups commonly employed in E2 reactions are listed in Table 2.1. As you can see, they are essentially the same as those displaced in nucleophilic substitution reactions (see Part 2), with two exceptions. First, protonated alcohols are not listed as substrates RX in Table 2.1, because they usually react by the E1 mechanism (as we shall see later) rather than the E2 mechanism. Secondly, the trimethylammonium and dimethylsulfonium groups have limited importance as leaving groups in substitution reactions, although they are particularly important in elimination reactions. In fact, the reaction involving trimethylammonium is known, after its discoverer, as the Hofmann elimination

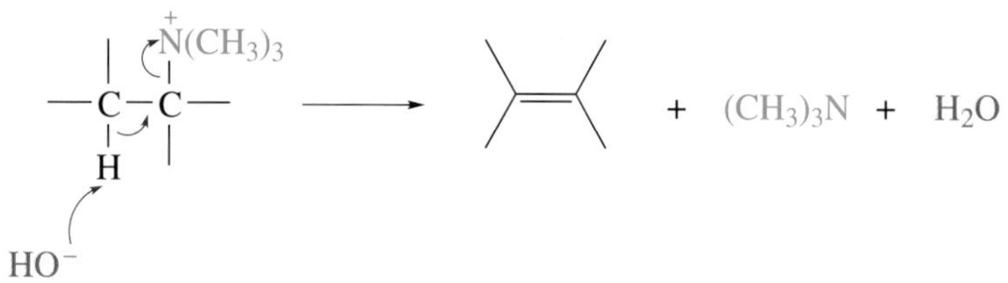

We shall discuss later how to make ammonium hydroxide salts such as R$_4$N$^+$ $^-$OH, which are known as *quaternary* ammonium salts because N has four alkyl groups attached to it, and which undergo Hofmann elimination reactions on heating.

In E2 reactions involving haloalkanes and alkyl tosylates, the base is added to the substrate to effect the reaction. The bases used most frequently are hydroxide (HO^-), alkoxide (RO^-) and amide ($^-NH_2$). (Note that the term 'amide' is used to describe both the carboxylic acid derivatives, $RCONH_2$, and the anion, $^-NH_2$. In the rest of Part 3, 'amide' will always refer to the anion $^-NH_2$.)

The other major factor determining the scope of the E2 reaction is the structure of the substrate. Primary, secondary *and tertiary* substrates can all react by the E2 mechanism (provided that there is a hydrogen atom on an adjacent carbon), although with primary and secondary substrates, the corresponding S_N2 reactions may compete effectively (tertiary substrates rarely react by the S_N2 mechanism).

Now that you have some idea of the scope of reactions proceeding by the E2 mechanism, let's go on to look at the stereochemistry of the process.

August Wilhelm von Hofmann (1818–1892)

August Hofmann (Figure 2.1) was born in Giessen and went to the local University in 1836 where he studied law and philosophy. He changed to chemistry after being inspired by Liebig. Having obtained his doctorate in 1841 he taught at the University of Bonn and in 1845 moved to London to become the Professor and Director of the Royal College of Chemistry (now Imperial College). In the early nineteenth century Britain had fallen behind Germany in experimental subjects such as chemistry, and so the Royal College of Chemistry was set up with the backing of Peel, Disraeli and Gladstone. Under Liebig's recommendation Prince Albert persuaded Hoffman to become the first Professor of Chemistry. He remained there for 20 years and trained many of the most influential British chemists, such as Perkin, who started the synthetic-dye industry in Britain. Hofmann returned to Germany to the University of Berlin in 1865, where he founded the German Chemical Society.

He is best known for his work on amines and ammonium compounds and there are several transformations that bear his name. He was one of the founders of the dyestuff industry and also first suggested the word 'valence'.

Figure 2.1
August Wilhelm von Hofmann (1818–1892).

2.2 The stereochemistry of the E2 mechanism

The stereochemistry of the E2 mechanism can be deduced by looking at a substrate in which both eliminated atoms or groups are attached to chiral carbon atoms, for example Compound **2.1**. Elimination of the hydrogen and bromine atoms from the *R,R*-Compound **2.1** will give Compound **2.2**

- We haven't drawn out the full structure of Compound **2.2**. How many alkenes can be represented by the structure shown?

- Certain alkenes can exist as stereoisomers. There are two possible isomers, *Z* (**2.3**) and *E* (**2.4**), represented by Structure **2.2**

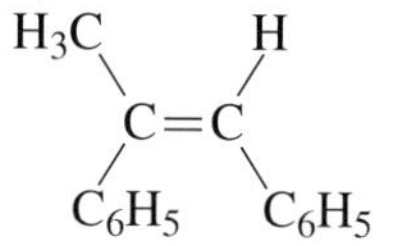

2.3
Z

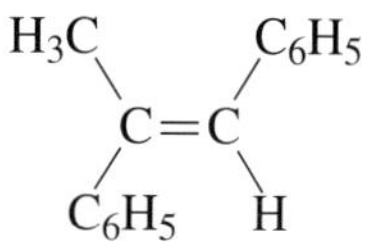

2.4
E

However, when Compound **2.1** is treated with base, only one alkene, Compound **2.3**, is formed. This is powerful evidence for a concerted one-step mechanism. A two-step mechanism would be expected to give a mixture of both alkenes.

This observation also tells us about the conformation adopted by Compound **2.1** during the elimination reaction. E2 eliminations occur either from the antiperiplanar conformation or from the synperiplanar conformation. This is because the movement of electrons, as revealed by the curly arrows, is more efficient when the H, C, C and X atoms all lie in the same plane

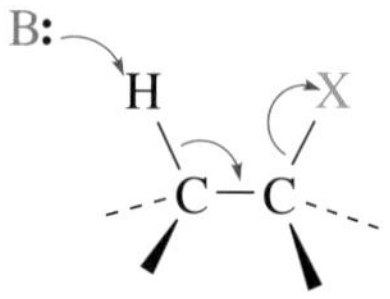

synperiplanar conformation

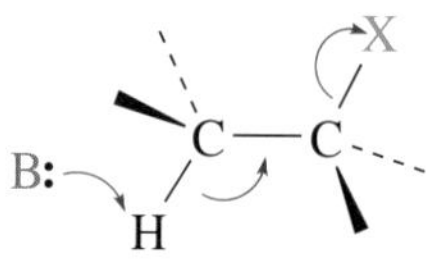

antiperiplanar conformation

Elimination from the antiperiplanar conformation, Structure **2.1a**, would give alkene Structure **2.3** (***anti*-elimination**). Elimination from the synperiplanar conformation, Structure **2.1b**, would give alkene Structure **2.4** (***syn*-elimination**). As only Compound **2.3** is formed from this reaction, *anti*-elimination must occur preferentially.

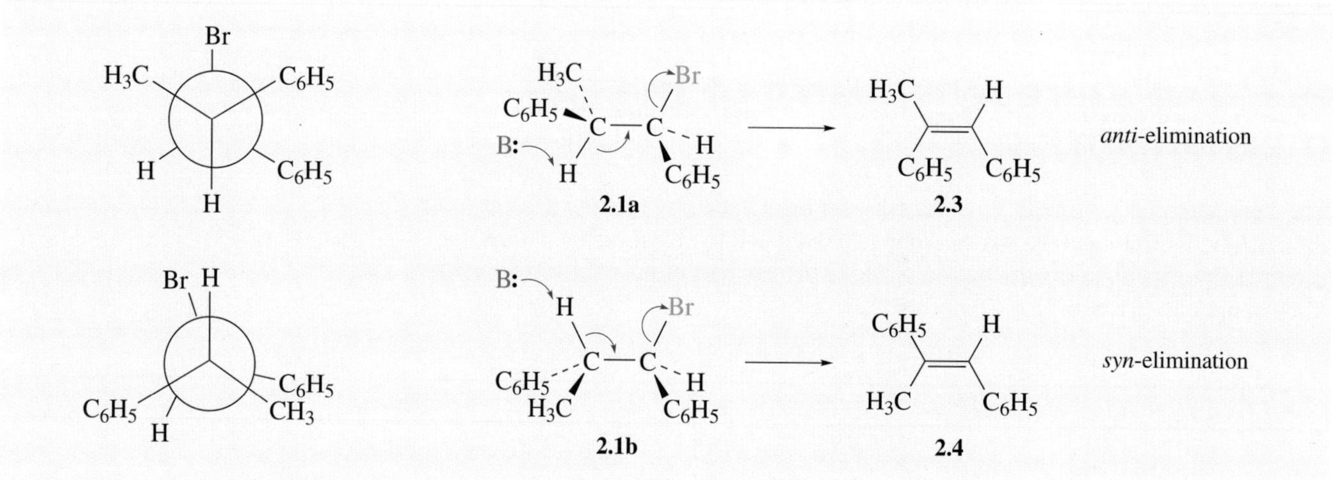

Figure 2.2 The Newman projections (on the left in each case) emphasize that the eclipsed form, Structure **2.1b**, will have a higher energy than the staggered form, Structure **2.1a**.

The antiperiplanar conformation Structure **2.la** is staggered (Figure 2.2), and therefore of lower energy than the eclipsed synperiplanar conformation, Structure **2.1b**. This energy difference will be reflected in the activated complexes for *syn*- and *anti*-elimination such that the energy of activation for *anti*-elimination will be less than that for *syn*-elimination, and the *anti*-process will proceed more quickly.

Another way of looking at this, which explains why *anti*-elimination is preferred from the antiperiplanar conformation, is that the electron pair from the breaking C—H bond displaces the bromide anion by backside attack. This keeps the incoming electrons and the outgoing electrons as far apart as possible. Just as backside attack by a nucleophile is favoured in the S_N2 mechanism, so backside displacement of the leaving group is favoured in the E2 elimination

S_N2 E2

All the E2 reactions discussed in this book are of the *anti*-elimination type, although some *syn*-eliminations are known.

COMPUTER ACTIVITY 2.1 Elimination exercise

At some point in the near future you should study the *Elimination exercise* in the CD-ROM that accompanies this book. This uses WebLab to examine the consequence of synperiplanar or antiperiplanar eliminations and thus allows you to predict the stereochemistry of the alkene formed in an elimination.

BOX 2.1 Ethene — the key to ripening

One way of speeding up the ripening of tomatoes is to place a banana in with the green tomatoes. This is because bananas give off ethene which is a plant-growth substance. Some plants are very sensitive to ethene, and levels as low as 1 part per million can cause ripening — in ripening apples the concentration of ethene is about 2 500 parts per million. This provides a way of controlling ripening during transport, where cylinders of ethene are employed. The plant growth properties of ethene are used in slow-release agrochemicals where elimination reactions form ethene at a fixed rate thereby promoting growth.

2.3 Isomeric alkenes in E2 reactions

In the majority of the reactions that we have seen so far, only one product has been possible by the *anti*-elimination mechanism. But this is not always so.

⬤ Look at Compounds **2.5** and **2.6**. Why can *anti*-elimination lead to more than one product for each compound?

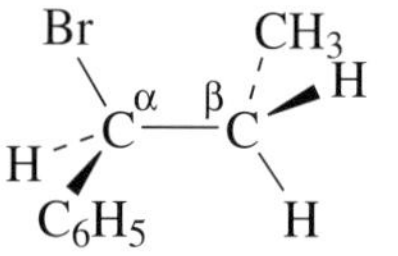

2.5

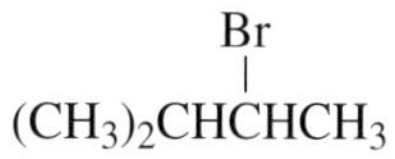

$(CH_3)_2CHCHCH_3$ with Br

2.6

⬤ In Compound **2.5** there are two hydrogens attached to the carbon β to the bromine (β-hydrogen atoms), and two alkenes, *E* (*trans*) and *Z* (*cis*), can be formed from different antiperiplanar conformations

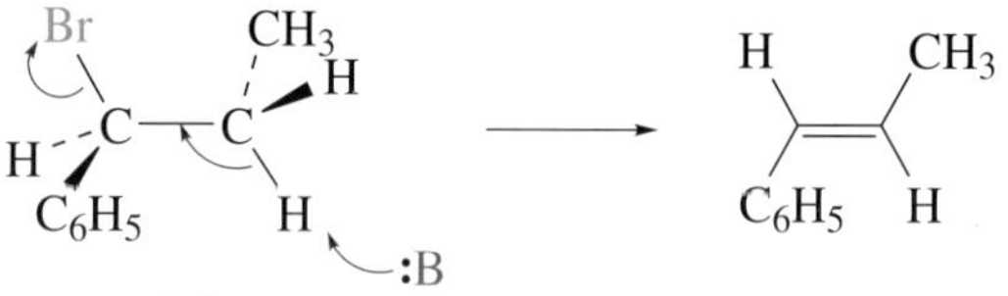

2.5   *trans*

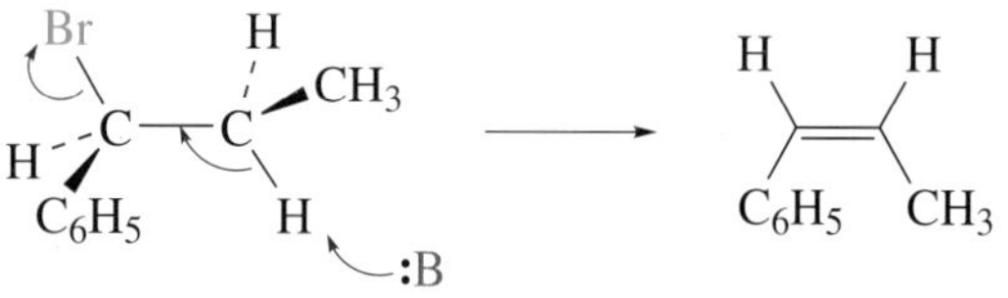

2.5   *cis*

In Compound **2.6** there are two β carbons, each with hydrogens attached

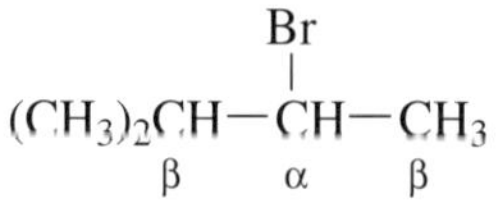

$$(CH_3)_2CH-CH-CH_3$$
$$\beta \quad \alpha \quad \beta$$

with Br on the α carbon

So elimination can occur in two directions

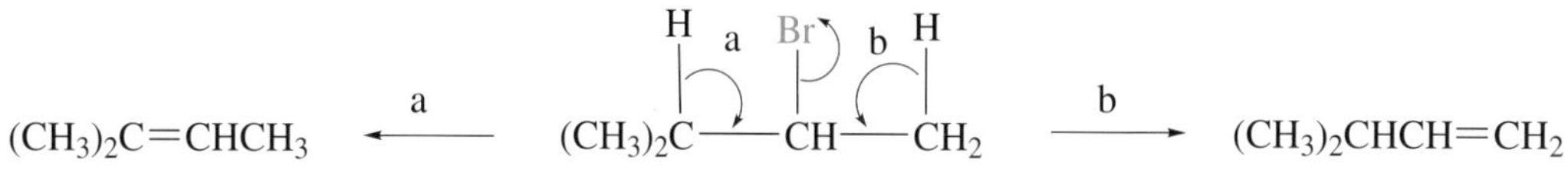

$$(CH_3)_2C=CHCH_3 \xleftarrow{\;a\;} (CH_3)_2C-CH-CH_2 \xrightarrow{\;b\;} (CH_3)_2CHCH=CH_2$$

2.6

One aim of organic chemists is to be able to predict the major product of a reaction with some degree of confidence. This is not always possible with E2 reactions when there is a choice of β-hydrogen atoms capable of attaining an antiperiplanar relationship to the leaving group. Usually, a mixture of all possible products is obtained, and for a particular substrate the ratio of the products is generally found to depend on the choice of base and reaction conditions. However, there are certain rules that have emerged from the study of a great number of elimination reactions. The application of these rules usually enables chemists to predict which product will predominate. Let's look at these rules.

2.3.1 Which isomer will predominate?

As we have seen, elimination from certain substrates by the antiperiplanar E2 mechanism can lead to two stereoisomers.

> As a rule, elimination occurs preferentially from the lower energy antiperiplanar conformation.

Look at the following example; we have drawn the two possible antiperiplanar Conformations **A** and **B**

Br   H
  \   /CH₂CH₃
H–C—C   —HBr→   H   H
  /CH₃   H   \   /
                           C=C
     A   H₃C   CH₂CH₃
                              cis

Br   CH₂CH₃
  \   /H
H–C—C   —HBr→   H   CH₂CH₃
  /CH₃   H   \   /
                           C=C
     B   H₃C   H
                              trans

- Is Conformation **A** or Conformation **B** of lower energy?

- This is perhaps best seen from the corresponding Newman projections

A   **B**

In Conformation **A**, the ethyl group lies between the bromine and the methyl group, whereas in Conformation **B** it lies between the bromine and a hydrogen. So Conformation **B** is of lower energy than Conformation **A** and therefore **B** is the preferred conformation.

Application of the above rule leads to the prediction that elimination will occur more readily from Conformation **B**. In practice this is found to be so; treatment of this substrate with potassium ethoxide, $K^{+-}OCH_2CH_3$, as base gives 51% of the *trans*-alkene (from Conformation **B**, Figure 2.3) and only 18% of the *cis*-alkene (from Conformation **A**).

- Elimination involving one of the hydrogens in the terminal methyl group could also occur. What is the product of this reaction?

- This third product, which is formed in 31% yield, is:

H   H
  \   /CH₂CH₂CH₃
H–C—C   —HBr→   H   H
  /H   Br   \   /
                           C=C
                          H   CH₂CH₂CH₃

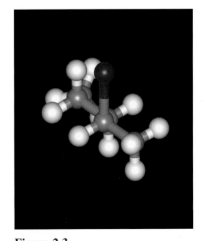

Figure 2.3
Ball and stick model of *anti*-elimination from the most stable (preferred) Conformation **B**.

2.3.2 Which direction of elimination?

By looking at a number of reactions in which two directions of elimination are possible, we can formulate empirical rules that enable us to predict which product will predominate. These rules apply only when the two alkene products differ in their **degree of substitution**. The degree of substitution is given by the number of substituents, other than hydrogen, attached to the double bond; for example

$RCH=CH_2$	monosubstituted
$R_2C=CH_2$ or $RCH=CHR$	disubstituted
$R_2C=CHR$	trisubstituted
$R_2C=CR_2$	tetrasubstituted

It is generally found that the thermodynamic stability of alkenes increases with increasing substitution. Some representative results from E2 reactions in which a more substituted and a less substituted alkene are formed are shown in Table 2.2.

Table 2.2 Product distributions from elimination reactions

(i) $\underset{|}{\overset{Br}{CH_3CH_2CHCH_3}}$ + $CH_3CH_2O^-$ ⟶ $CH_3CH=CHCH_3$ + $CH_3CH_2CH=CH_2$
  80% 20%

(ii) $\underset{|}{\overset{Br}{CH_3CH_2CH_2CHCH_3}}$ + $CH_3CH_2O^-$ ⟶ $CH_3CH_2CH=CHCH_3$ + $CH_3CH_2CH_2CH=CH_2$
  69% 31%

(iii) $\underset{|}{\overset{I}{CH_3CH_2CH_2CHCH_3}}$ + $CH_3CH_2O^-$ ⟶ $CH_3CH_2CH=CHCH_3$ + $CH_3CH_2CH_2CH=CH_2$
  80% 20%

(iv) $\underset{|}{\overset{\overset{+}{N}(CH_3)_3\ Br^-}{CH_3CH_2CHCH_3}}$ + HO^- ⟶ $CH_3CH=CHCH_3$ + $CH_3CH_2CH=CH_2$
  5% 95%

(v) $\underset{|}{\overset{\overset{+}{S}(CH_3)_2\ Br^-}{CH_3CH_2CHCH_3}}$ + $CH_3CH_2O^-$ ⟶ $CH_3CH=CHCH_3$ + $CH_3CH_2CH=CH_2$
  26% 74%

● Look at the examples in Table 2.2. In which cases do the more substituted alkenes predominate, and in which cases do the less substituted alkenes predominate?

● In examples (i), (ii) and (iii) the more substituted alkenes predominate. In examples (iv) and (v) the less substituted alkenes predominate.

The data in Table 2.2 are representative of many elimination reactions. The product distribution is found to depend on whether the substrate is neutral (with a halide or sulfonate leaving group) or *positively charged* (with a $-\overset{+}{N}(CH_3)_3$ or $-\overset{+}{S}(CH_3)_2$ leaving group — which become $N(CH_3)_3$ and $S(CH_3)_2$ respectively). The empirical rules linking the nature of the substrate and the product distribution are known as **Saytzev's rule** and **Hofmann's rule**.

Saytzev's rule

Neutral substrates give predominantly the most substituted alkene (that is, the thermodynamically most stable one), although appreciable amounts of other possible alkenes are usually obtained (Table 2.2, examples (i), (ii) and (iii)).

Hofmann's rule

Positively charged substrates show a strong tendency to form predominantly the least substituted alkene (Table 2.2, examples (iv) and (v)).

The reasons for this remarkable dependence on the nature of the leaving group are still not fully understood. However, a knowledge of these rules can be exploited to great effect. For example, the same bromoalkane (Structure **2.7**) can be used to prepare two related alkene isomers in good yield

$$\underset{\textbf{2.7}}{\underset{|}{\overset{Br}{\overset{|}{CH_3CH_2CHCH_3}}}} \xrightarrow{CH_3CH_2O^-} \underset{\text{but-2-ene (major product)}}{CH_3CH=CHCH_3}$$

Most substituted

$(CH_3)_3N \Big|$

$$\underset{\textbf{2.8}}{\underset{|}{\overset{\overset{+}{N}(CH_3)_3\ Br^-}{\overset{|}{CH_3CH_2CHCH_3}}}} \xrightarrow{HO^-} \underset{\text{but-1-ene (major product)}}{CH_3CH_2CH=CH_2}$$

Least substituted

Notice that the bromoalkane (Structure **2.7**) can be transformed into the quaternary ammonium bromide (Structure **2.8**) by treatment with trimethylamine by an S_N2 reaction. Treatment of **2.8** with hydroxide anion forms the quaternary ammonium hydroxide, which undergoes elimination on heating.

- Which product results from Saytzev elimination, and which from Hofmann elimination?

- But-2-ene (the more substituted alkene) is produced from the neutral substrate by Saytzev elimination, whereas but-1-ene is produced from the charged substrate, by Hofmann elimination.

Alexander Mikhaylovich Saytzev (1841–1910)

Alexander Saytzev was born in Kazan in Russia on 20 June 1841. He studied in Germany at Marburg and Leipzig and with Butlerov at Kazan. In fact both Saytzev and Markovnikov were students of Butlerov. Saytzev became Professor of Chemistry at the University of Kazan, where he worked on the synthesis of alcohols from esters, ketones and aldehydes using zinc and alkyl halides. His student Reformatsky continued his work. Saytzev's brother Mikhayl was also a chemist and ran the chemical works in Kazan.

Read the following summary of the rules for predicting the major product of an
E2 elimination then apply these rules in answering Questions 2.1 and 2.2.

Rules for predicting the major product of an E2 elimination

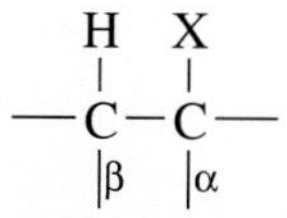

1 The substrate must bear a hydrogen atom in a β-position relative to a
suitable leaving group

2 Elimination will occur preferentially from an antiperiplanar conformation
such as that shown in Structure **2.9**

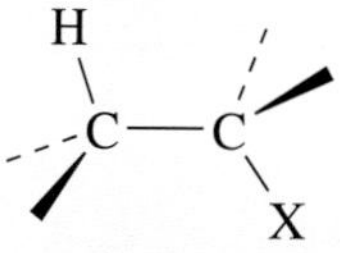

3 If there is a choice of β-hydrogen atoms, the following rules should be
applied:

(a) If both β-hydrogen atoms are on the same carbon atom, elimination
occurs preferentially from the lower-energy antiperiplanar (preferred)
conformation.

(b) If the β-hydrogen atoms are on different carbon atoms, and the
substrate is neutral, the most substituted alkene predominates
(Saytzev's rule). If the substrate is positively charged, the least
substituted alkene predominates (Hofmann's rule).

It may be necessary to apply both (a) and (b) to predict the major product of
an E2 elimination.

2.9

QUESTION 2.1

Compound **2.10** is a diastereomer of Compound **2.1**. What elimination
product or products will be formed from Compound **2.10** in a bimolecular
mechanism? Use Newman projections or ISIS Draw and WebLab ViewerLite
to look at the conformations if necessary.

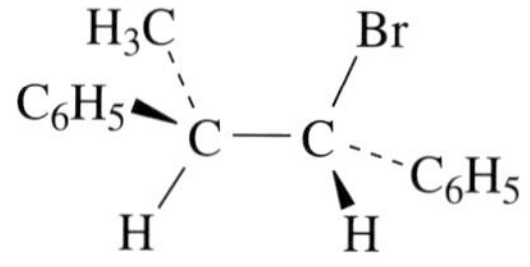

2.10

QUESTION 2.2

List all the products that could result when the following substrates (i)–(iv)
react by an *anti*-E2 mechanism. Say which will be the major product from
each substrate, and give your reasons. Newman projections will be essential
here, and ISIS Draw/WebLab may also be useful.

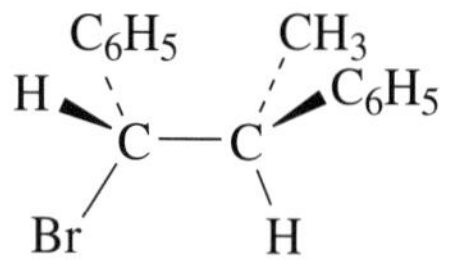

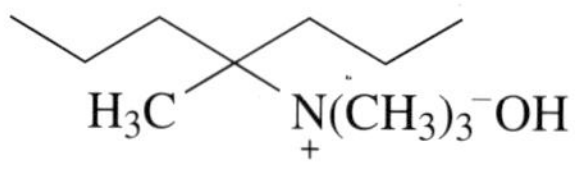

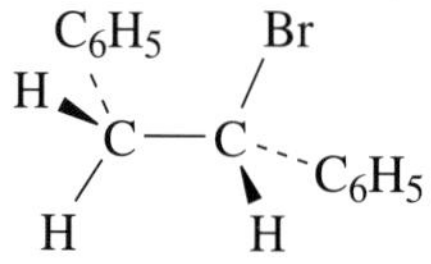

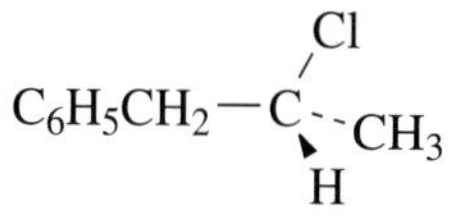

Each of the organic reactants and products in (i) to (iv) above is available in
WebLab ViewerLite on the CD-ROM associated with this book.

3

THE E1 MECHANISM

In the E1 mechanism the leaving group departs *before the proton is removed* in a subsequent step. In common with the S_N1 reaction, the first step is rate-limiting, and results in the formation of an intermediate carbocation. Elimination, to form the neutral alkene, results if the carbocation is then deprotonated by a base

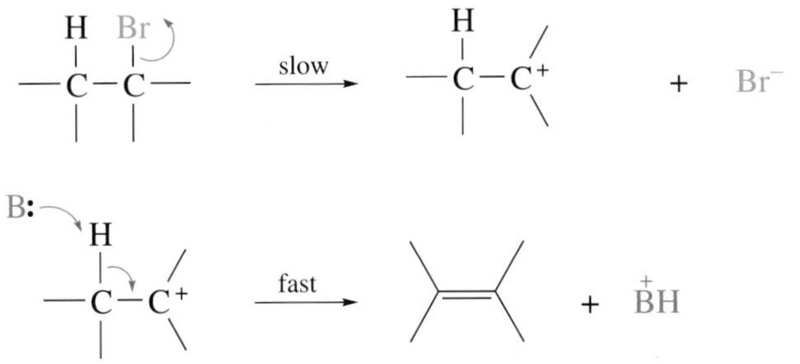

● How would you expect the rate of the E1 reaction to vary with the concentration of base?

● The rate-limiting step is the first step, which is a unimolecular process, and so the rate should depend only on the substrate concentration. Variations in base concentration, therefore, have no effect on the rate of the E1 mechanism.

In fact, E1 reactions usually occur in the absence of a strong base. They are normally carried out by heating the substrate in a solvent that can solvate the ions formed in the slow step; such solvents are usually sufficiently basic to remove the proton in the subsequent fast step

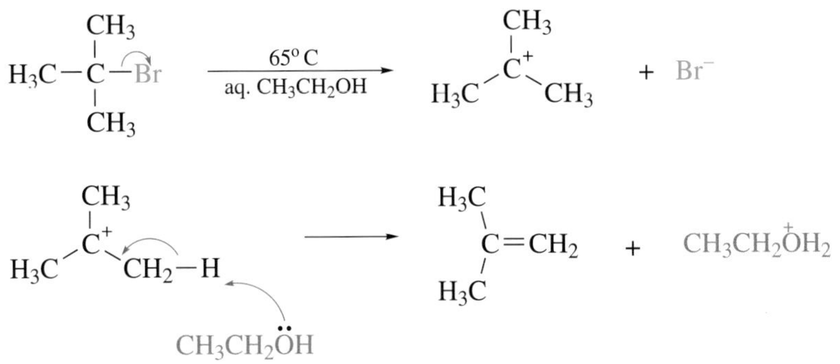

You may wonder why the Br^- does not act as the base in this reaction. In the reaction there are many more ethanol molecules than Br^- ions. So the ethanol is the more likely to be present to remove the proton from the carbocation as it forms.

In order for a substrate to react by the E1 mechanism, it must lead to a relatively stable carbocation intermediate.

● Primary, secondary and tertiary substrates can all react by the E2 mechanism. Which are more likely to react by the E1 mechanism?

● Tertiary and then secondary substrates are more likely to react by the E1 mechanism, because tertiary and secondary carbocations are relatively stable. Primary substrates rarely react by the E1 mechanism.

So the structure of the substrate does much to determine the mechanism of the elimination reaction.

● Do you think that the nature of the leaving group is as important in determining the distribution of products in the E1 mechanism as it is in the E2 mechanism?

● In the E2 mechanism the nature of the leaving group often determines which product will predominate — Saytzev or Hofmann. In the E1 mechanism the leaving group is expelled in the first step, before the double bond is formed, and not surprisingly it has little influence over the products of the reaction.

● What alkene products would Structures **3.1** and **3.2** lead to if they were to react by an E2 mechanism? Would the same products be obtained via an E1 mechanism?

$$\underset{\textbf{3.1}}{\overset{\overset{\displaystyle Br}{|}}{(CH_3)_2CCH_2CH_3}} \qquad\qquad \underset{\textbf{3.2}}{\overset{\overset{\displaystyle \overset{+}{S}(CH_3)_2\ Br^-}{|}}{(CH_3)_2CCH_2CH_3}}$$

● If elimination were to proceed by the E2 mechanism, Compound **3.1** would be expected to obey Saytzev's rule and give the more substituted alkene (Structure **3.3**), whereas Compound **3.2** should give predominantly the less substituted product (Structure **3.4**) according to Hofmann's rule.

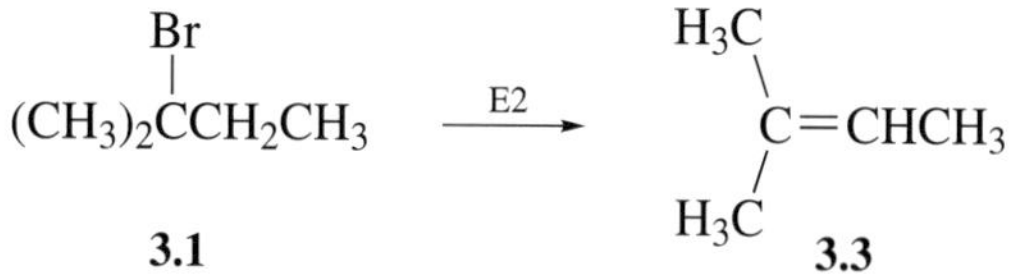

$$\underset{\textbf{3.1}}{\overset{\overset{\displaystyle Br}{|}}{(CH_3)_2CCH_2CH_3}} \xrightarrow{\ E2\ } \underset{\textbf{3.3}}{\overset{\displaystyle H_3C}{\underset{\displaystyle H_3C}{>}}C=CHCH_3}$$

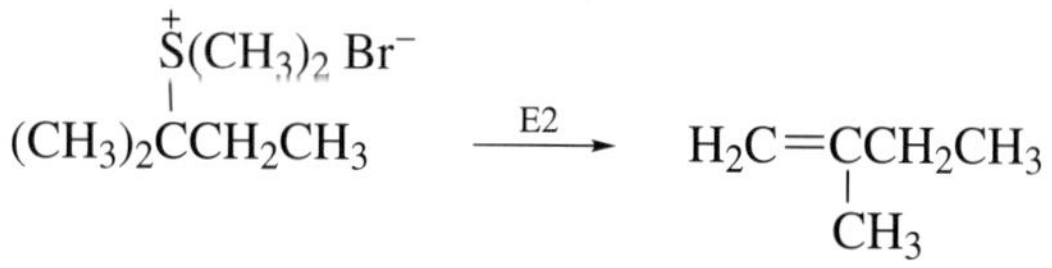

$$\underset{\textbf{3.2}}{\overset{\overset{\displaystyle \overset{+}{S}(CH_3)_2\ Br^-}{|}}{(CH_3)_2CCH_2CH_3}} \xrightarrow{\ E2\ } \underset{\textbf{3.4}}{\overset{\displaystyle H_2C=CCH_2CH_3}{\underset{\displaystyle CH_3}{|}}}$$

However, in the E1 mechanism, Compounds **3.1** and **3.2** form the same carbocation intermediate, and so in this case you might expect a similar product distribution from both substrates.

This is indeed the experimental result, and the predominant alkene formed from **3.1** and **3.2** by an E1 mechanism is **3.3**, the more substituted alkene (that is, the more thermodynamically stable for both substrates). E1 reactions are generally found to obey Saytzev's rule, whatever the leaving group

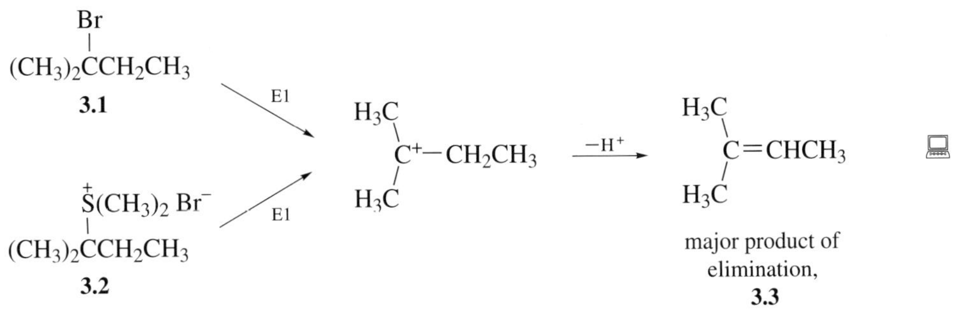

Br
|
(CH$_3$)$_2$CCH$_2$CH$_3$

3.1

$\xrightarrow{\text{E1}}$

$\overset{+}{\text{S}}$(CH$_3$)$_2$ Br$^-$
|
(CH$_3$)$_2$CCH$_2$CH$_3$

3.2

$\xrightarrow{\text{E1}}$

H$_3$C
\
C$^+$—CH$_2$CH$_3$
/
H$_3$C

$\xrightarrow{-\text{H}^+}$

H$_3$C
\
C=CHCH$_3$
/
H$_3$C

major product of
elimination,
3.3

Since the most stable alkene is formed in E1 eliminations, when diastereomers are possible, the more stable *E (trans)* form usually predominates over the *Z (cis)* form

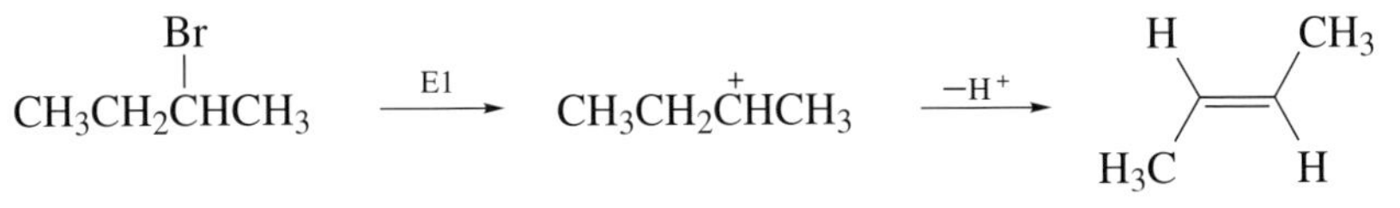

Br
|
CH$_3$CH$_2$CHCH$_3$ $\xrightarrow{\text{E1}}$ CH$_3$CH$_2$$\overset{+}{\text{C}}HCH_3$ $\xrightarrow{-\text{H}^+}$

H   CH$_3$
\   /
C=C
/   \
H$_3$C   H

E (trans)

Alcohols usually undergo elimination by the E1 mechanism, and this reaction is very useful. Acid catalysis is required to convert the hydroxyl group into a good leaving group; one possible mechanism is shown in Reaction 3.1. Note that although this is an acid-catalysed reaction, a base is still needed to remove a proton and complete this elimination. If the acid used is sulfuric acid then the hydrogen sulfate anion (HSO$_4{}^-$) probably acts as the base, but the solvent can also act as the base, as we saw in the E1 elimination of hydrogen bromide.

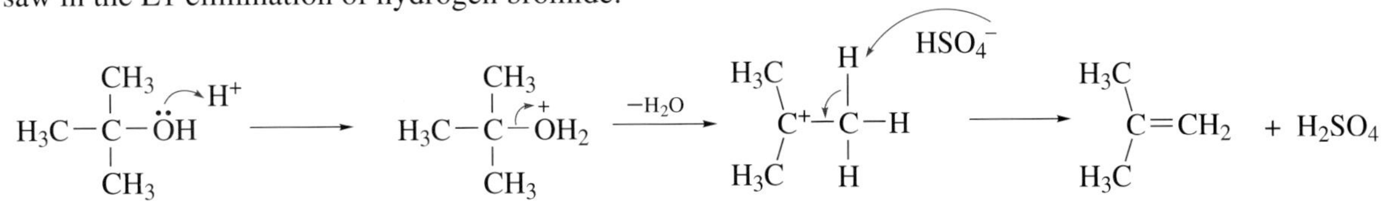

CH$_3$
|
H$_3$C—C—$\ddot{\text{O}}$H $\overset{\frown}{}H^+$
|
CH$_3$

$\longrightarrow$

CH$_3$
|
H$_3$C—C—$\overset{+}{\text{O}}$H$_2$
|
CH$_3$

$\xrightarrow{-\text{H}_2\text{O}}$

H$_3$C   H   HSO$_4^-$
\ | ⌒
C$^+$—C—H
/ |
H$_3$C   H

$\longrightarrow$

H$_3$C
\
C=CH$_2$ + H$_2$SO$_4$ (3.1)
/
H$_3$C

In general, however, the E1 reaction is rarely used for preparative purposes because, as we shall see in the next section, a mixture of elimination and substitution products is usually obtained.

In the next section we shall discuss the competition between substitution and elimination reactions. To understand this competition it is important that you have acquired a good grasp of the mechanisms of substitution and elimination reactions.

BOX 3.1 Insect pheromones

Insect pheromones are organic compounds that are used by insects to communicate with each other. There are sex, trail, alarm and defence pheromones that are given off by insects to inform other insects of opportunities and problems. Pheromones have been exploited by 'baiting' insect traps with an appropriate sex pheromone, which lures a specific insect to its doom. Many of these pheromones are alkenes, often with Z (*cis*) double bonds. A selection is shown below

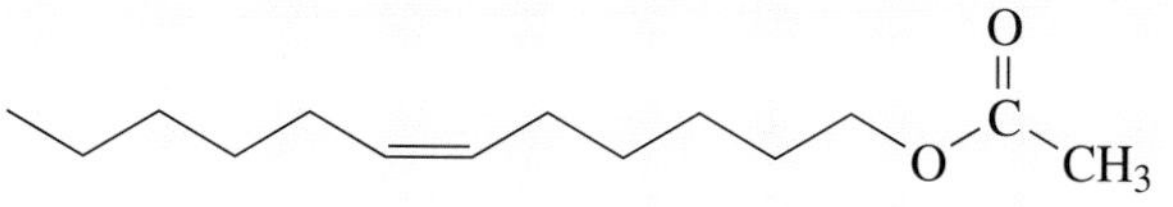

Z-dodec-6-en-1-yl ethanoate
(female sugar beet moth)

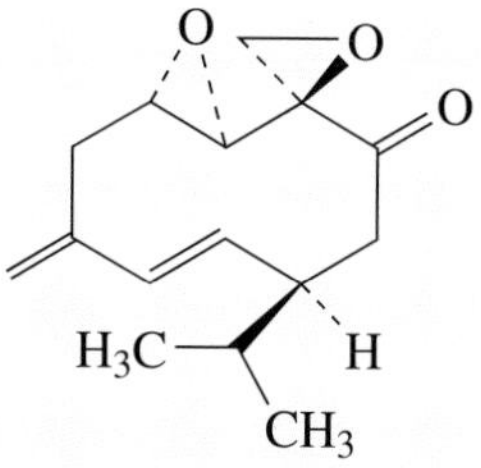

American cockroach

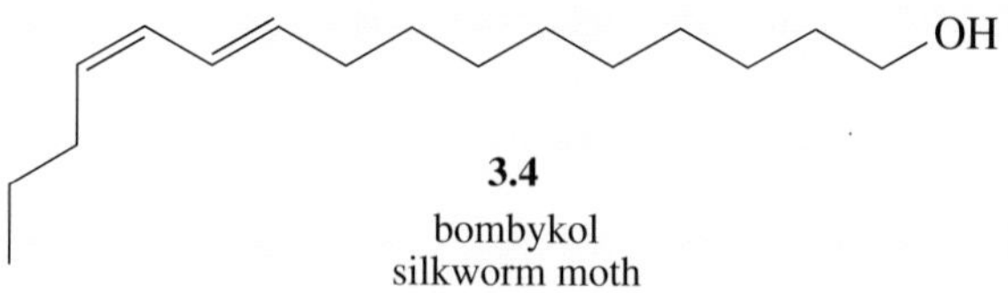

3.4
bombykol
silkworm moth

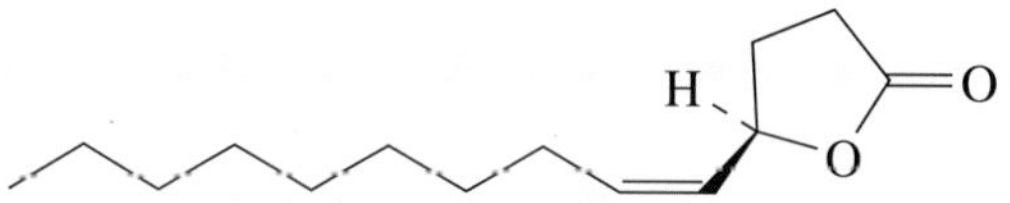

Japanese beetle

Insects can detect very small amounts of materials and the activity often depends upon the stereochemistry of the alkene. For example, bombykol (Structure **3.4**), the sex attractant for the male silkworm moth, has a *trans–cis* configuration, which is 10 billion times more active than the *cis–trans* form and 10 trillion times more active than the *trans–trans* compound. The exact synthesis of such compounds is therefore important if we wish to lure the insects into traps and thus keep them away from crops. Thus, the synthetic chemist needs to be aware of the stereochemistry of reactions such as eliminations to ensure they form the active product in the highest yield.

3.1 Summary of Sections 2 and 3

1 The leaving groups in the bimolecular elimination mechanism (E2) are essentially the same as those in substitution reactions (that is, halides and tosylate). Alcohols do not usually react by the E2 mechanism, but quaternary ammonium salts, such as $R\overset{+}{N}(CH_3)_3X^-$ and sulfonium salts, such as $R\overset{+}{S}(CH_3)_2X^-$ do.

2 Hydroxide ($^-$OH), alkoxide ($^-$OR) and amide($^-$NH$_2$) anions are the most commonly used bases in the E2 reaction.

3 Primary, secondary *and tertiary* substrates can all react by the E2 mechanism.

4 E2 elimination occurs preferentially from a conformation in which the eliminated groups are opposed to each other in an antiperiplanar conformation. This process is known as *anti*-elimination.

5   If there are two hydrogen atoms on the same β-carbon atom, both of which can undergo *anti*-elimination, elimination occurs preferentially from the lowest energy antiperiplanar conformation.

6   With some substrates, elimination can occur in two directions, to give different alkene products. Saytzev's rule states that neutral substrates give predominantly the *more* substituted alkene, by the E2 mechanism. Hofmann's rule applies to charged substrates, such as sulfonium and ammonium salts, and these give a predominance of the *less* substituted alkene by the bimolecular mechanism.

7   The two-step E1 mechanism usually occurs with tertiary and secondary substrates in the absence of a strong base. The leaving group has little effect on the product distribution and, where applicable, the more substituted Saytzev products are formed.

8   Reactions proceeding by the E1 mechanism are not usually used to prepare alkenes.

QUESTION 3.1

Without turning back to earlier sections depict the (a) El, (b) E2, (c) S_N1 and (d) S_N2 mechanisms using curly arrows. *Briefly* outline the main points of each mechanism, and draw an energy profile for each.

QUESTION 3.2

Reaction 3.1

$$H_3C-\underset{\underset{CH_3}{|}}{\overset{\overset{CH_3}{|}}{C}}-\ddot{O}H \xleftarrow{} H^+ \longrightarrow H_3C-\underset{\underset{CH_3}{|}}{\overset{\overset{CH_3}{|}}{C}}\overset{+}{\ddot{O}}H_2 \xrightarrow{-H_2O} \underset{H_3C}{\overset{H_3C}{>}}C^+ \underset{H}{\overset{H}{\underset{|}{C}}}-H \quad HSO_4^- \longrightarrow \underset{H_3C}{\overset{H_3C}{>}}C=CH_2 \; + \; H_2SO_4$$

shows the development of a charged leaving group. What will be the predominant elimination product when

$$CH_3-CH_2-\underset{\underset{\textstyle OH}{|}}{CH}-CH_3$$

is treated with H_2SO_4?

QUESTION 3.3

Give the mechanism and products for the reaction of Structure **3.5** (i) with sodium ethoxide via an E2 mechanism and (ii) with water via an E1 mechanism.

$$CH_3-CH_2-CH_2-\underset{\underset{\textstyle N(CH_3)_3}{\overset{\textstyle +}{|}}}{CH}-CH_3$$

3.5

QUESTION 3.4

How would you prepare **3.6** via an E2 mechanism?

$$CH_3-CH_2-CH_2-CH=CH-CH_2-CH_3$$

3.6

ELIMINATION VERSUS SUBSTITUTION

4

So far in this book we have considered substitution and elimination reactions in isolation. But in fact the majority of substrates that can undergo substitution can also undergo elimination; all that is required is a β-hydrogen atom capable of attaining the required orientation to the leaving group, for example

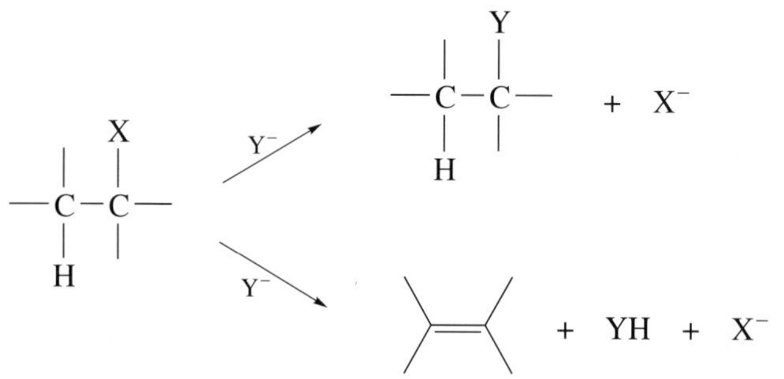

We have dealt with substitution and elimination separately to enable you to appreciate the different factors involved. But in practice most reactions of this type give a mixture of substitution and elimination products, for example

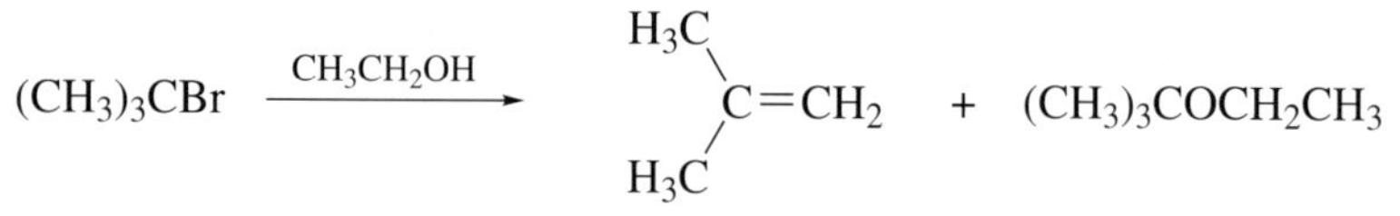

What is required from a preparative viewpoint is a means of obtaining 100% elimination when an alkene is required, and 100% substitution when the substituted product is required. In practice this ideal situation can rarely be obtained, but there are guidelines that can be applied to *maximize* the yield of the desired product. A detailed discussion of the factors affecting the ratio of elimination to substitution is beyond the scope of this book. What we will do is discuss *some* of the more important points, and leave you with the set of rules that practising chemists apply when planning a reaction.

4.1 Substrate structure

One important factor in determining the ratio of substitution to elimination is the structure of the substrate. This factor, however, is not usually a variable: the structure of the desired product determines the structure of the substrate. So let's leave the substrate structure, and discuss factors that apply to all types of substrate.

4.1.1 Unimolecular versus bimolecular mechanism

The main factor that can be exploited is the mechanism of the reaction. If the reaction conditions are varied, either a bimolecular or a unimolecular mechanism can be favoured. This means that the competition is between E2 and S_N2, or between E1 and S_N1. Generally, bimolecular mechanisms are of greater preparative value. This is because unimolecular mechanisms proceed by way of the same carbocation intermediate, whereas bimolecular mechanisms follow totally different pathways

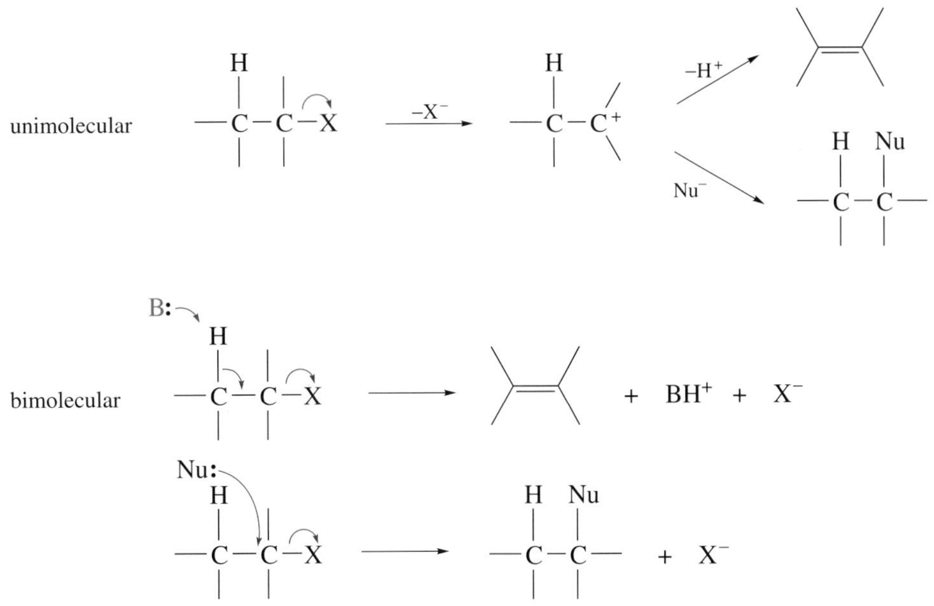

Because of the common intermediate in S_N1 and E1 reactions, it is usually difficult to influence the product ratio in unimolecular reactions to any great extent.

Think back to Part 2. Do you remember the factors favouring bimolecular mechanisms?

A high concentration of a reactive reagent.

As we shall see for bimolecular reactions, by controlling the nature of the reagent, the leaving group and the temperature, we can control whether substitution or elimination predominates.

4.2 Choice of reagent and other factors

If we assume that a bimolecular mechanism will operate for the reaction of a given substrate, then it is the choice of *reagent* that largely decides the ratio of substitution to elimination. To encourage *substitution,* a high concentration of a nucleophile that is not a good base should be used (Cl^- or ^-CN). To encourage *elimination,* a high concentration of a good base should be used. However, many good bases (for example HO^-, CH_3O^-) are also good nucleophiles, and when E2 and S_N2 mechanisms are in competition, a mixture of elimination and substitution products results. To maximize the yield of alkene under these circumstances a *non-nucleophilic base* can be used. The best example of such a compound is potassium tertiary-butoxide, Structure **4.1**, abbreviated to potassium *tert*-butoxide (Figure 4.1). The *tert*-butoxide anion is so bulky that it cannot form a bond to carbon in an S_N2 mechanism (it is a poor nucleophile), but it can abstract a proton (it is a good base).

$$K^{+\,-}O-\underset{\underset{CH_3}{|}}{\overset{\overset{CH_3}{|}}{C}}-CH_3$$

4.1

So halides, which normally give a mixture of substitution and elimination products, form only alkenes with potassium *tert*-butoxide.

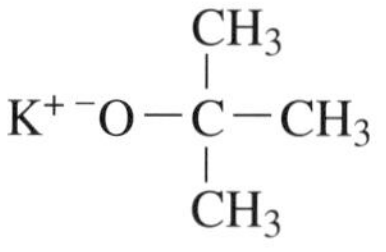

4.2.1 Choice of leaving group

In general, neutral leaving groups (Br, Cl, etc.) favour substitution, whereas positively charged leaving groups ($\overset{+}{N}(CH_3)_3$ and $\overset{+}{S}(CH_3)_2$) favour elimination, as can be seen from the following data.

Table 4.1 How the leaving group favours either substitution or elimination

$$C_2H_5X \xrightarrow{\ \ C_2H_5O^-\ \ } H_2C{=}CH_2 \quad + \quad C_2H_5OC_2H_5$$

X = Br	1%	99%
X = $\overset{+}{S}(CH_3)_2$	80%	20%
X = $\overset{+}{N}(CH_3)_3$	70%	30%

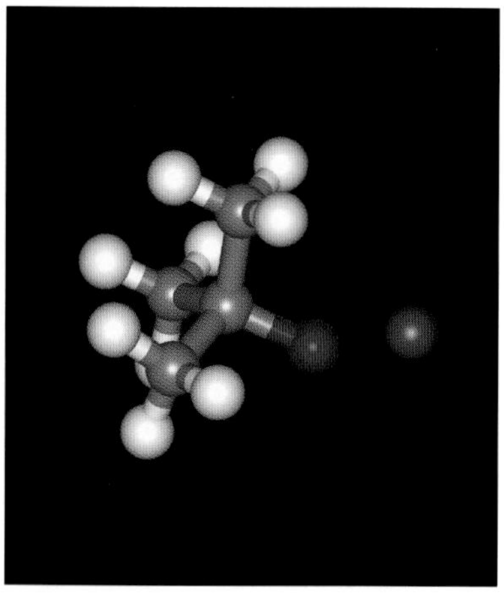

Figure 4.1
Ball and stick model of potassium tertiary–butoxide.

4.2.2 Temperature

In general, a high temperature favours elimination, as can be seen from the following data.

Table 4.2 A high reaction temperature favours elimination products

$$(CH_3)_2CHBr \xrightarrow{\ \ HO^-\ \ } (CH_3)_2CHOH \quad + \quad CH_3CH{=}CH_2$$

45° C	47%	53%
100° C	36%	64%

4.2.3 Summing up

With secondary and primary substrates, the yield of *substitution* products (by the S_N2 mechanism) can be maximized by using

- a nucleophile that is not a good base (such as Cl^- or ^-CN),
- a low reaction temperature,
- a neutral leaving group.

However, with nucleophiles that are good bases (HO^- and RO^-) some elimination will also occur. This problem is compounded when tertiary substrates are employed, because, although they can react by an E2 mechanism, they rarely react by an S_N2 mechanism, for steric reasons. So elimination usually predominates with a strong base or nucleophile

$$(CH_3)_3CBr \xrightarrow{CH_3CH_2O^-} \begin{matrix} H_3C \\ \diagdown \\ C=CH_2 \\ \diagup \\ H_3C \end{matrix} + (CH_3)_3COCH_2CH_3$$

E2 83% S_N1 17%

In situations like this, the unimolecular pathway often gives a better yield of the substitution product if a weak base is employed

$$(CH_3)_3CBr \xrightarrow{CH_3CH_2OH} \begin{matrix} H_3C \\ \diagdown \\ C=CH_2 \\ \diagup \\ H_3C \end{matrix} + (CH_3)_3COCH_2CH_3$$

E1 19% S_N1 81%

4.3 Summary of Section 4

1   Elimination and substitution reactions usually occur in competition.

2   Bimolecular mechanisms are usually preferable from a preparative viewpoint because the yield of the desired product can usually be maximized by application of the following rules.

For elimination: (i) use a high concentration of a good base; (ii) if necessary use a base that is a poor nucleophile, for example $(CH_3)_3CO^-$ (iii) use a charged leaving group ($^+S(CH_3)_2$ or $^+N(CH_3)_3$); (iv) use a high reaction temperature.

For substitution: (i) use a high concentration of a good nucleophile; (ii) use a neutral leaving group (Br, etc.); (iii) use a low reaction temperature. If the desired nucleophile is also a good base, especially for tertiary substrates, an unfavourable mixture of products will probably result. A unimolecular process may then be better.

QUESTION 4.1

For each of the reactions (i)–(iv) give the structures of the possible substitution and elimination products. Which type of reaction (S_N or E) do you think will predominate?

(i) $(CH_3)_2CHCHBrCH_3$ $\xrightarrow{NC^-}$

(ii) $(CH_3)_2CHCHBrCH_3$ $\xrightarrow{CH_3CH_2O^-}$

(iii) $(CH_3)_2CHCHBrCH_3$ $\xrightarrow{(CH_3)_3CO^-}$

(iv)   ⬡—$CH_2OSO_2C_6H_4CH_3$ $\xrightarrow{KI}$

Each of the organic reactants and products in (i) to (iv) above is available in WebLab ViewerLite on the CD-ROM associated with this book.

QUESTION 4.2

Compound **4.2** can undergo an elimination by treating with sulfuric acid or by converting it into a *para*-toluenesulfonate ester, Structure **4.3**, followed by treatment with base. Which route would enable us to maximize the yield of elimination product?

$$CH_3-CH_2-\overset{\overset{\displaystyle OH}{|}}{CH}-CH_2-CH_3$$
4.2

$$CH_3-CH_2-\overset{\overset{\displaystyle OSO_2C_6H_4CH_3}{|}}{CH}-CH_2-CH_3$$
4.3

OTHER USEFUL ELIMINATION REACTIONS

5

At the start of Part 3, we said that the general form of a β-elimination reaction is as follows

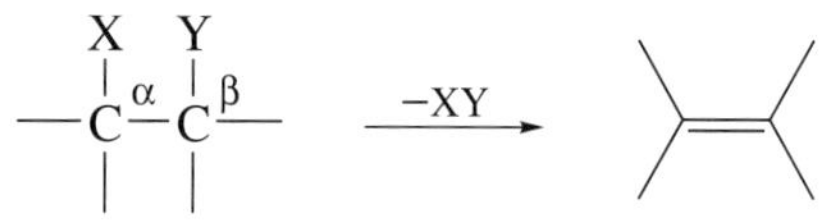

We then went on to discuss elimination reactions in which Y is H. However, other β-elimination reactions are known in which neither X nor Y is a hydrogen atom. These reactions are useful because the position of the new double bond is defined by the position of X and Y, and also because elimination usually occurs to the exclusion of substitution. Let's look at two such examples, and then go on to consider how alkynes are produced by elimination reactions.

5.1 Dehalogenation and decarboxylative elimination

1,2-dihalides can easily be converted into the corresponding alkenes by treatment with either zinc dust or sodium iodide. The reaction, called **dehalogenation**, proceeds by the one-step E2 mechanism, with elimination generally occurring preferentially from an antiperiplanar conformation, Reaction 5.1.

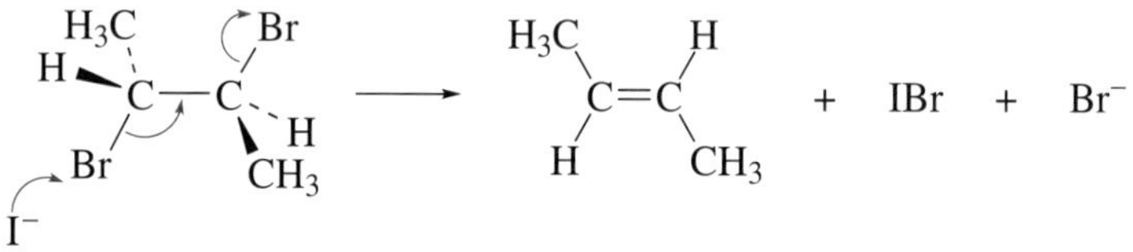

Notice that both zinc and iodide act as nucleophiles, forming a new bond to one of the halogen atoms. The electrons in the broken carbon–halogen bond form the double bond. Note also that the iodide could act as a nucleophile and substitute one or both of the bromine atoms. However, in 1,2-dihalide systems of this type the elimination product predominates.

A related elimination reaction occurs when β-halocarboxylic acids, especially those that have two substituents on the carbon atom next to the acid group, are treated with base

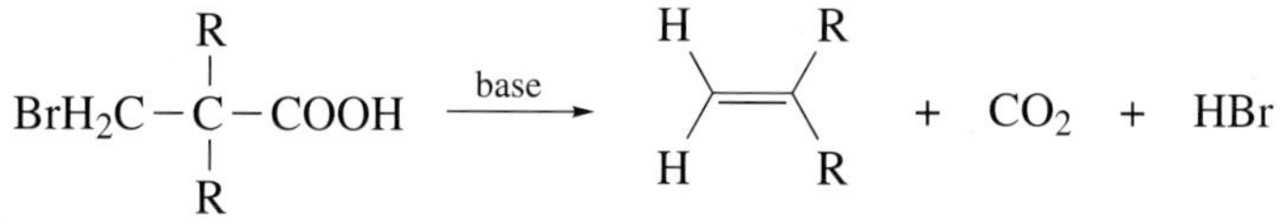

○ What other reaction could occur in unsubstituted β-halocarboxylic acids (that is, as above but with a hydrogen atom in place of each R group)?

$$BrH_2C-\underset{\underset{H}{|}}{\overset{\overset{H}{|}}{C}}-COOH$$

5.1

○ Normal elimination of hydrogen and bromine would then be possible, to give propenoic acid, Structure **5.1**.

At first sight, this type of reaction might seem unrelated to those that we have looked at previously, although if you think of the whole carboxylic acid group (COOH) as one of the groups to be eliminated you can see that it is on the carbon atom adjacent to the bromine leaving group, as below. So let's think about the mechanism.

$$-\underset{\underset{|}{\overset{\overset{Br}{|}}{C}}}{}-\underset{\underset{|}{\overset{\overset{COOH}{|}}{C}}}{}-$$

○ What will happen first when a carboxylic acid is treated with a base?

○ The acid will form a salt with the base by the transfer of a proton, thereby forming a carboxylate anion.

$$BrH_2C-\overset{CH_3}{\underset{CH_3}{\overset{|}{C}}}-C\overset{O}{\underset{O-H}{\diagdown}} \longrightarrow Br-CH_2-\overset{CH_3}{\underset{CH_3}{\overset{|}{C}}}-C\overset{O}{\underset{O^-}{\diagdown}} \longrightarrow H_2C=\overset{CH_3}{\underset{CH_3}{\overset{|}{C}}} + O=C=O + Br^- \quad (5.3)$$

$$HO^-$$

$$+ \quad H_2O$$

The carboxylate anion then eliminates carbon dioxide and a bromide anion, to form the alkene product (Reaction 5.3). This type of reaction, called **decarboxylative elimination**, is sometimes important in biological systems.

5.2 Preparation of alkynes by elimination reactions

Elimination reactions can also be used to prepare alkynes. Dihalides can be converted into alkynes by treatment with base, provided that the two halogen atoms are on the same, or adjacent, carbon atoms. In general, alkyne formation requires more drastic reaction conditions than alkene formation, and a strong base, such as sodium amide, is usually employed (Reactions 5.4 and 5.5). The product of Reaction 5.5, phenylethyne, is shown in Figure 5.1 as a ball and stick model.

$$CH_3CH_2CH_2CH_2CHCl_2 \xrightarrow{NaNH_2} CH_3CH_2CH_2CH=CHCl \xrightarrow{NaNH_2} CH_3CH_2CH_2C\equiv CH \quad (5.4)$$

an alkenyl halide

$$C_6H_5\underset{\overset{|}{Br}}{CH}CH_2Br \xrightarrow{NaNH_2} \left\{ \begin{array}{c} C_6H_5\overset{\overset{Br}{|}}{C}=CH_2 \\ \text{or} \\ C_6H_5CH=CHBr \end{array} \right\} \xrightarrow{NaNH_2} C_6H_5C\equiv CH \quad (5.5)$$

These reactions proceed by way of a alkenyl halide (that is, an alkene carrying a halogen substituent). The strong base usually ensures that the E2 mechanism operates.

Alkenes can be converted into 1,2-dihalides by treatment with chlorine or bromine. So the addition-elimination process provides a useful way of transforming alkenes into alkynes

$$RCH{=}CH_2 \xrightarrow{X_2} \underset{\displaystyle RCHCH_2X}{\overset{\displaystyle X}{|}} \xrightarrow[\text{base}]{\text{strong}} RC{\equiv}CH$$

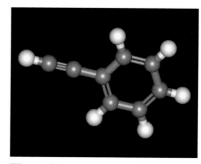

Figure 5.1
Ball and stick model of phenylethyne, notice the linear nature of the alkyne bond.

5.3 Summary of Section 5

1   In dehalogenation and decarboxylative elimination reactions, the double bond is formed in a defined position.

2   The dehalogenation of 1,2-dihalides can be carried out using zinc or sodium iodide. The reaction occurs by an E2 mechanism, and so the halogen substituents must be able to adopt an antiperiplanar relationship.

3   On treatment with base, β-halocarboxylic acids undergo decarboxylative elimination to give alkenes.

4   Elimination reactions involving dihalides can be used to prepare alkynes. A strong base such as sodium amide is usually required.

QUESTION 5.1

Predict the products of the following elimination reactions. Use curly arrows to illustrate the reaction mechanisms. What is the stereochemistry of the product in (a)?

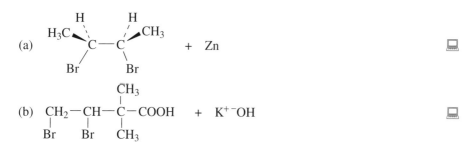

LEARNING OUTCOMES FOR PART 3

Now that you have completed Part 3 *Elimination: Pathways and Products*, you should be able to do the following:

1 Recognize valid definitions of and use in a correct context the terms, concepts and principles in the following table (All questions).

List of scientific terms, concepts and principles used in Part 3 of this book.

Term	Page no	Term	Page no
anti-elimination	196	E1 mechanism	192
base	191	E2 mechanism	191
β-elimination	189	elimination reaction	189
decarboxylative elimination	213	Hofmann's rule	201
degree of substitution	200	Saytzev's rule	201
dehalogenation	212	*syn*-elimination	196

2 Use curly arrows to illustrate the course of an elimination reaction proceeding by a given mechanism (E1 or E2). (Questions 1.1, 3.3 and 5.1)

3 Given the substrate and the mechanism of an elimination reaction:

(a) Predict the structure of the product(s), indicating the stereochemistry where necessary;

(b) Predict which elimination product will predominate, where more than one product can be formed. (Questions 2.1, 2.2, 3.2, 3.3, 3.4 and 5.1)

4 Depict the El, E2, S_N1 and S_N2 mechanisms using both curly arrows and reaction coordinate diagrams, and outline the differences between these mechanisms. (Question 3.1)

5 Predict the possible elimination and substitution products from a given reaction, and predict which product is likely to predominate. (Questions 4.1 and 4.2)

6 Provide a mechanistic explanation for a given stereochemical observation relating to an elimination reaction. (Questions 2.1, 2.2 and 5.1)

QUESTIONS: ANSWERS AND COMMENTS

QUESTION 1.1 (Learning Outcome 2)

(a)

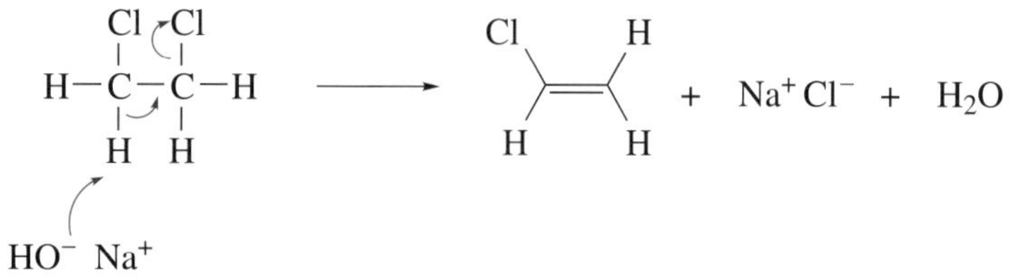

Note that the arrow from the base (hydroxide ion) starts on the negative charge (which signifies an electron pair), and shows a bond being formed to a hydrogen atom. The second arrow shows the formation of the new carbon–carbon bond: note that it starts in the centre of the original C—H bond, and that it terminates in the centre of the bond being formed. The third arrow starts in the centre of the bond to be broken (C—Cl), and ends on the chlorine atom. The chloride anion is paired with the sodium cation in the product. A similar sequence of events occurs in (b).

(b)

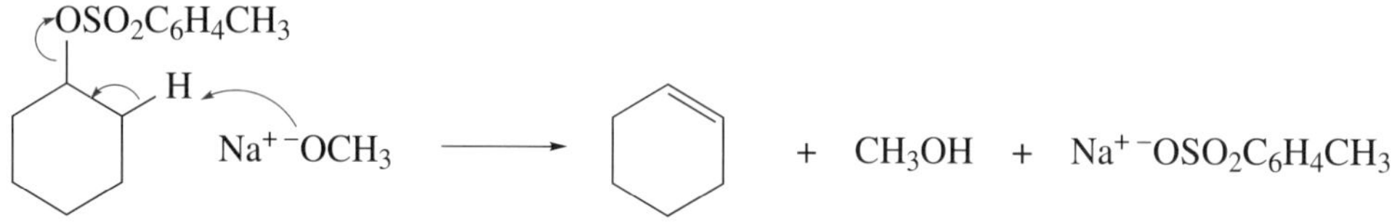

The methoxide anion acts as a base and removes a proton from the carbon β to the tosylate group. At the same time the tosylate group is lost and a double bond is formed.

(c)

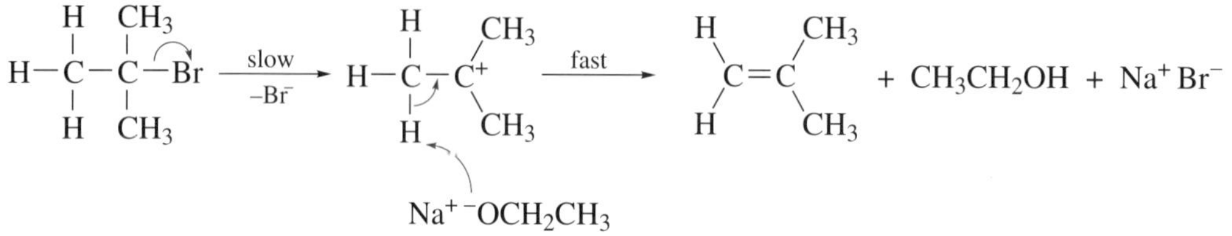

In (c) the first step is slow ionization, that is the C—Br bond breaks, to form the carbocation intermediate. Then, in a fast step, the ethoxide ion removes a proton from one of the carbon atoms adjacent to the carbon atom with the positive charge and the double bond forms. Notice that in this cation, removal of a hydrogen from any of the three adjacent carbon atoms gives the same alkene. Check that your curly arrows start and finish in the same places as those in the diagrams.

QUESTION 2.1 (Learning Outcomes 3 and 6)

Compound **2.10** can exist in an infinite number of conformations, but the conformation depicted is the one necessary for *anti*-elimination. The product of this reaction will have a *trans*-arrangement with respect to the two phenyl groups (that is, Compound **2.4**)

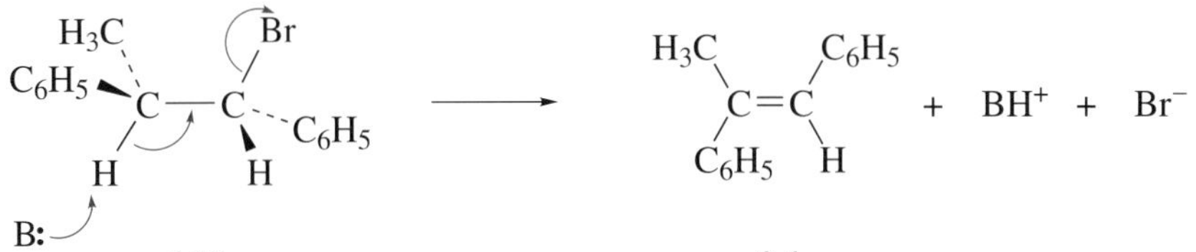

2.10 → **2.4**

QUESTION 2.2 (Learning Outcomes 3 and 6)

(i) Only one product is possible by an *anti*-E2 mechanism. Elimination would not occur from the conformation shown in the question, but rather from the alternative conformation, which results from internal rotation

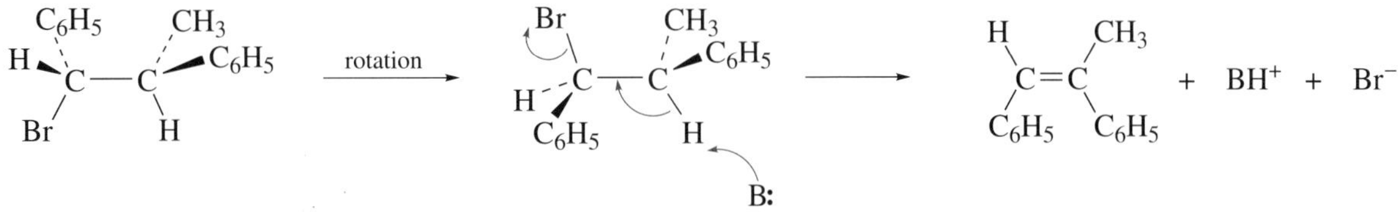

rotation

(ii) There are two β-hydrogen atoms capable of attaining an *anti*-relationship with the bromine atom giving either **S.1** or **S.2**.

(a)

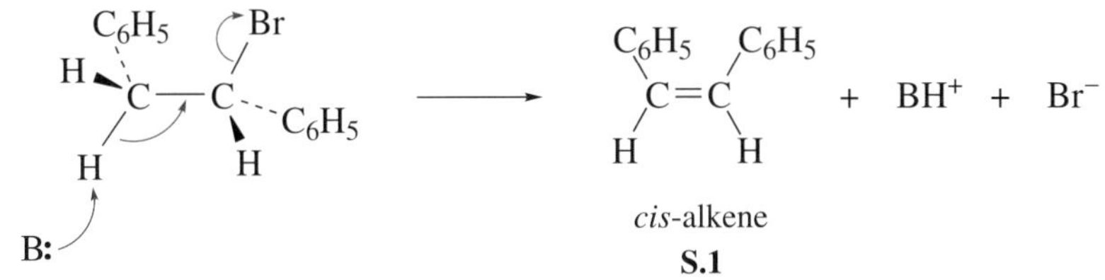

cis-alkene
S.1

(b)

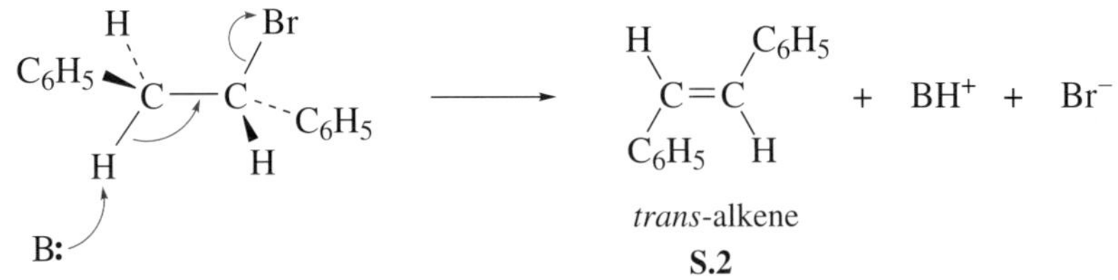

trans-alkene
S.2

Process (a) proceeds by the more crowded conformation (see below), and so the *trans*-alkene produced by process (b), would be expected to predominate. Use ISIS Draw and WebLab ViewerLite to look at the conformations if the Newman projections aren't helpful.

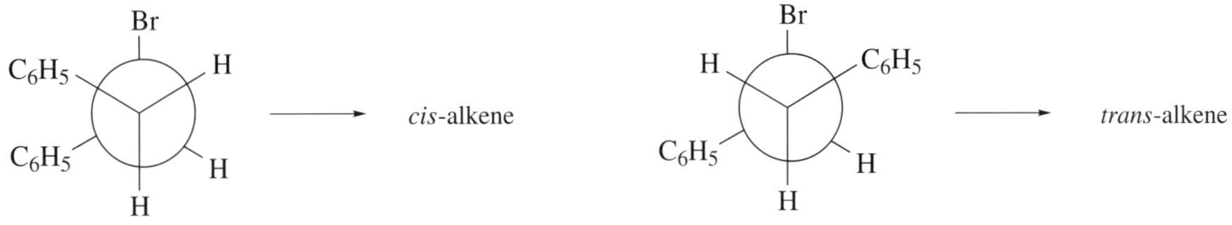

cis-alkene

trans-alkene

(a)

(b)

(iii) Hofmann's rule applies here, because the leaving group is charged. So the least substituted alkene would be expected to predominate

$+ \ H_2O \ + \ N(CH_3)_3$

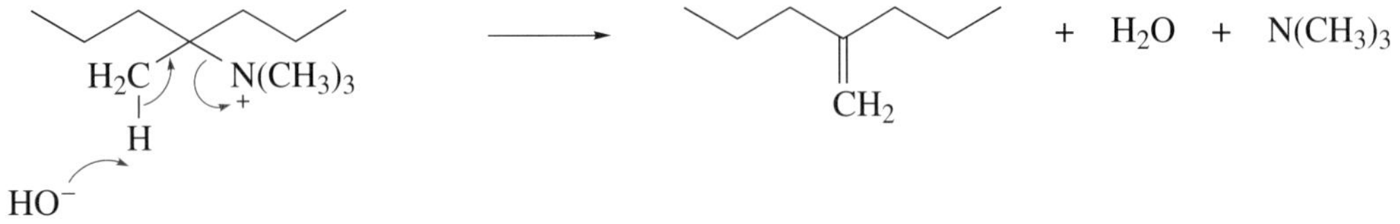

Two minor products (E and Z isomers, shown below) arise from Saytzev elimination.

Z-isomer   *E*-isomer

(iv) Three elimination products, **S.3**, **S.4** and **S.5** are possible from this reaction, and all three are likely to be formed to some extent. However, as the leaving group is neutral, Saytzev's rule operates, and so the more substituted products **S.3** and **S.4** would be expected to predominate.

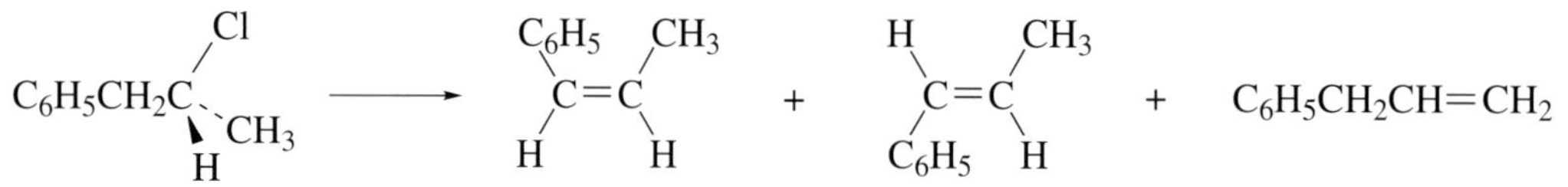

$C_6H_5CH_2C$ (Cl, CH_3, H) $\longrightarrow$

C_6H_5 CH_3 $C=C$ H H	H CH_3 $C=C$ C_6H_5 H	$C_6H_5CH_2CH=CH_2$
S.3	**S.4**	**S.5**

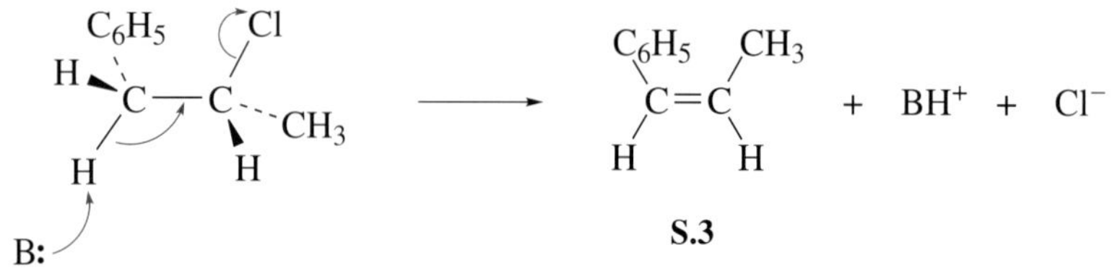

B: → (mechanism) $\longrightarrow$

C_6H_5 CH_3 $C=C$ H H + BH^+ + Cl^-

S.3

B: → (mechanism) $\longrightarrow$

H CH_3 $C=C$ C_6H_5 H + BH^+ + Cl^-

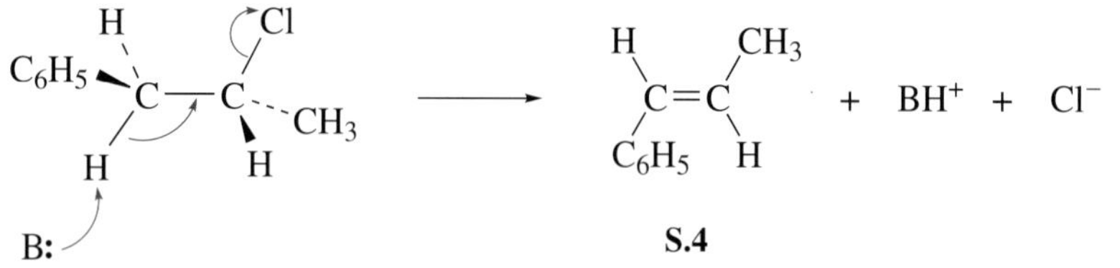

S.4

By looking at the three-dimensional structures (draw the Newman projections or use ISIS Draw and WebLab ViewerLite if necessary) you should be able to see that the conformation leading to **S.4** is of lower energy than that leading to **S.3**. Hence, you should have predicted **S.4** as the major product. There is another feature worth pointing out here. **S.3** and **S.4** both have the new double bond next to the phenyl group, whereas **S.5** does not. Such an arrangement of alternating single and double bonds also reduces the energy of **S.3** and **S.4**, which, again, makes them preferred products.

QUESTION 3.1 *(Learning Outcome 4)*

(a) E1 mechanism

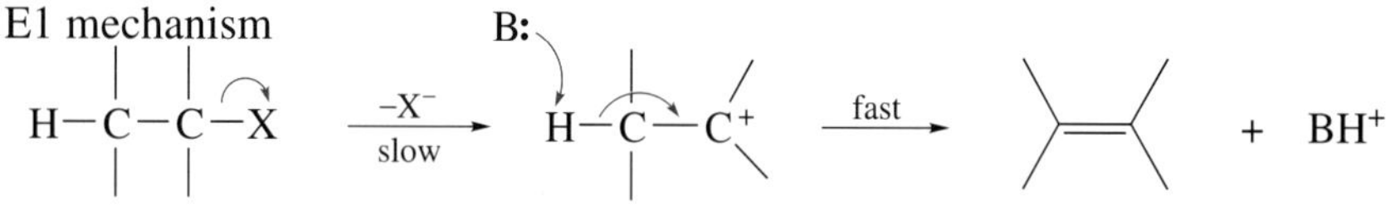

$H-C-C-X \xrightarrow[\text{slow}]{-X^-} H-C-C^+ \xrightarrow{\text{fast}} \;\;\diagdown C=C \diagup\; + \; BH^+$

Tertiary is more likely than secondary, and secondary is *much* more likely than primary to react by this mechanism. In brief

tertiary > secondary >> primary

$J = k[\text{substrate}]$

The energy profile for the E1 mechanism is shown in Figure Q.1.

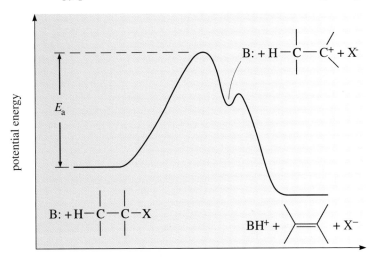

Figure Q.1 Energy profile for the E1 mechanism.

(b) E2 mechanism

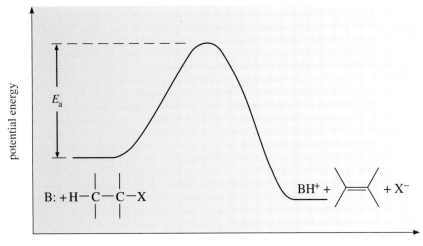

Primary, secondary and tertiary substrates can all react by this mechanism.

$J = k[\text{base}][\text{substrate}]$

Anti-elimination is favoured.

The energy profile for the E2 mechanism is shown in Figure Q.2.

Figure Q.2 Energy profile for the E2 mechanism.

(c) S_N1 mechanism

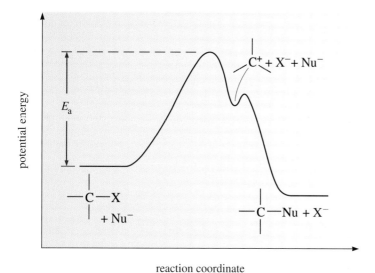

Tertiary is more likely than secondary, and secondary is much more likely than primary to react by this mechanism. In brief

tertiary > secondary >> primary

$J = k$[substrate]

Racemization of chiral substrates occurs.

The energy profile for the S_N1 mechanism is shown in Figure Q.3.

Figure Q.3 Energy profile for the S_N1 mechanism.

(d) S_N2 mechanism

Primary is more likely than secondary to react by this mechanism: this mechanism does not occur with tertiary substrates.

$J = k$[Nu$^-$][substrate]

Inversion of configuration occurs.

The energy profile for the S_N2 mechanism is shown in Figure Q.4.

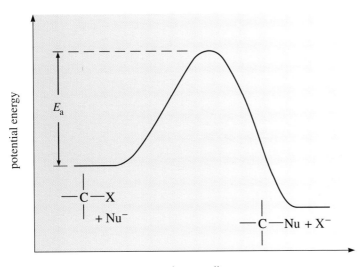

Figure Q.4 Energy profile for the S_N2 mechanism.

The mechanism of elimination is as follows

$$:\ddot{O}H \quad \curvearrow H^+$$
$$CH_3CH_2CHCH_3 \longrightarrow$$

$$\overset{+}{O}H_2$$
$$CH_3CH_2CHCH_3 \quad \xrightarrow{\text{slow}}$$

$$CH_3CH \overset{+}{-} \overset{+}{C}HCH_3$$
$$\underset{H}{\overset{}{|}}$$

$$HSO_4^-$$

$$CH_3CH{=}CHCH_3 \quad + \quad H_2SO_4$$

Although a charged leaving group is formed, it is lost before the position of the double bond is determined. As with other E1 eliminations, the more stable form will predominate, that is the more substituted *trans*-alkene.

(i) Reaction with sodium ethoxide via an E2 mechanism is shown below

$$\overset{+}{N}(CH_3)_3$$
$$CH_3{-}CH_2{-}CH_2{-}CH{-}CH_2 \longrightarrow CH_3{-}CH_2{-}CH_2{-}CH{=}CH_2 \ + \ CH_3{-}CH_2{-}OH$$
$$\underset{H}{\overset{}{|}}$$

$$CH_3{-}CH_2{-}O^-$$

Hoffman's rule suggests the least substituted alkene will be formed because of the charged leaving group.

(ii) Reaction with water via an E1 mechanism is shown below

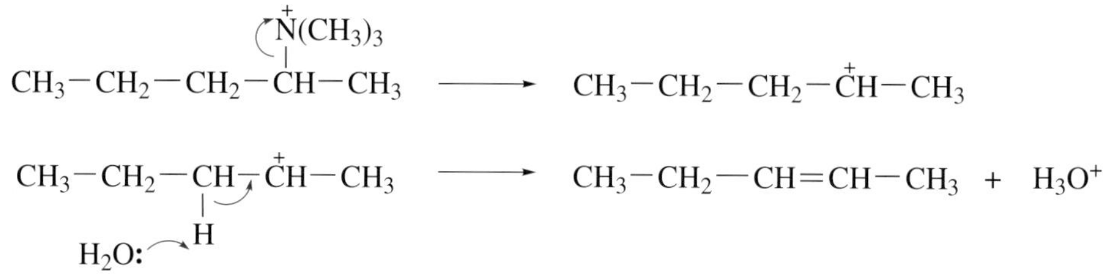

$$CH_3-CH_2-CH_2-\overset{\overset{+}{N}(CH_3)_3}{\underset{|}{CH}}-CH_3 \longrightarrow CH_3-CH_2-CH_2-\overset{+}{C}H-CH_3$$

$$CH_3-CH_2-\underset{\underset{|}{H_2O:}\overset{}{\frown}}{CH}\overset{+}{\frown}\overset{+}{C}H-CH_3 \longrightarrow CH_3-CH_2-CH=CH-CH_3 + H_3O^+$$

E1 reactions usually obey Saytzev's rule to give the more substituted alkene. Since the more stable alkene is formed, the *E* form predominates.

QUESTION 3.4 (Learning Outcome 3)

Compound **3.6** can be made by elimination of HX in two ways (where X is a halide or a tosylate group)

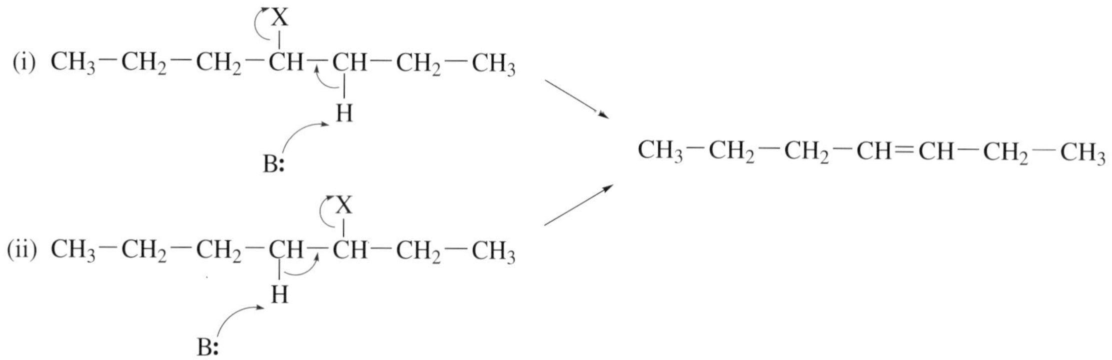

(i) $CH_3-CH_2-CH_2-\overset{\overset{X}{\frown}}{\underset{|}{CH}}-\overset{}{\underset{\underset{|}{H}}{CH}}-CH_2-CH_3$

          B:

         $CH_3-CH_2-CH_2-CH=CH-CH_2-CH_3$

(ii) $CH_3-CH_2-CH_2-\overset{\overset{X}{\frown}}{\underset{\underset{|}{H}}{CH}}-CH-CH_2-CH_3$

      B:

    S.6

Route (i) is favoured because the starting material is symmetrical and removal of a hydrogen from a carbon on either side of the C—X gives the same product. However, a competing elimination to route (ii) is route (iii)

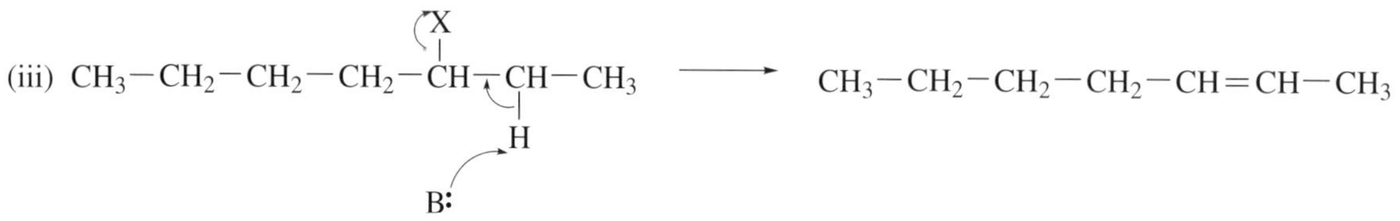

(iii) $CH_3-CH_2-CH_2-CH_2-\overset{\overset{X}{\frown}}{\underset{\underset{|}{H}}{CH}}-CH-CH_3 \longrightarrow CH_3-CH_2-CH_2-CH_2-CH=CH-CH_3$

      B:

  S.6

Since Structure **S.6** can lead to two alkenes, which are both disubstituted (routes (ii) and (iii)), we cannot predict which is formed predominantly and it is likely that they are formed in almost equal amounts.

QUESTION 4.1 (Learning Outcome 5)

(i) The cyanide ion is a very good nucleophile and a poor base. With a secondary halide as substrate, substitution would be expected to be the major (if not the only) reaction, occurring predominantly by an S_N2 mechanism, to produce compound **S.7**.

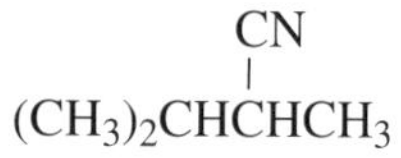

$$(CH_3)_2CH\overset{\overset{CN}{|}}{CH}CH_3$$

S.7

The less likely elimination product is Structure **S.8**.

(ii) The ethoxide ion is a good base as well as a good nucleophile. E2 and S$_N$2 mechanisms would both be expected to operate, and so the alkenes **S.8** and **S.9**, and the ether **S.10**, should be formed. Of the two possible alkenes, the more substituted (**S.8**) should predominate, according to Saytzev's rule. Note that as the carbon attached to the bromo group is chiral, inversion will take place on substitution.

$(CH_3)_2C{=}CHCH_3$ $(CH_3)_2CHCH{=}CH_2$ $\overset{\displaystyle OCH_2CH_3}{(CH_3)_2CH\overset{|}{C}HCH_3}$

S.8   **S.9**   **S.10**

(iii) The *tert*-butoxide ion is a strong base but a poor nucleophile. Substitution would not be expected, and the only reaction products should be Compounds **S.8** and **S.9**. Once again, **S.8** should predominate, according to Saytzev's rule.

(iv) Substitution by an S$_N$2 mechanism to give Compound **S.11** will be the major pathway. The substrate is primary, and iodide ion is a very good nucleophile and a poor base.

The less likely elimination product is **S.12**.

QUESTION 4.2 (Learning Outcome 5)

Section 4 emphasized the need to ensure a bimolecular mechanism if we are to control the amount of elimination versus substitution (and vice versa). Reaction of Compound **4.2** with sulfuric acid proceeds via an E1 mechanism

S.11

S.12

$$CH_3{-}CH_2{-}\overset{\displaystyle OH}{\overset{|}{C}H}{-}CH_2{-}CH_3 \xrightarrow{H^+} CH_3{-}CH_2{-}\overset{\displaystyle \overset{+}{O}H_2}{\overset{|}{C}H}{-}CH_2{-}CH_3$$

$$CH_3{-}CH_2{-}\overset{\displaystyle \overset{+}{O}H_2}{\overset{|}{C}H}{-}CH_2{-}CH_3 \xrightarrow{-H_2O} CH_3{-}CH_2{-}\overset{+}{C}H{-}CH_2{-}CH_3$$

$$CH_3{-}CH{-}\overset{+}{C}H{-}CH_2{-}CH_3 \longrightarrow CH_3{-}CH{=}CH{-}CH_2{-}CH_3$$
B:

However, reaction of base with the tosylate will proceed via an E2 elimination

$$CH_3{-}CH{-}\overset{\displaystyle OSO_2C_6H_4CH_3}{\overset{|}{C}H}{-}CH_2{-}CH_3 \longrightarrow CH_3{-}CH{=}CH{-}CH_2{-}CH_3$$
B:

Thus, elimination can be favoured in this E2 process by using a non-nucleophilic strong base, such as *tert*-butoxide.

However, concentrated sulfuric acid is the most frequently used dehydrating agent for elimination. This is because:

(a) it is much cheaper than tosyl chloride or *tert*-butoxide so it can be used in excess,

(b) it protonates ⁻OH groups and initiates loss of H_2O,

(c) H_2SO_4 is a very poor nucleophile but provides an adequate base.

QUESTION 5.1 (Learning Outcomes 2, 3 and 6)

(a) This reaction proceeds by a concerted E2 mechanism, which requires the eliminated groups to adopt an *anti*-relationship. The correct conformation is shown in Reaction S.1, and the only product is *trans*-but-2-ene.

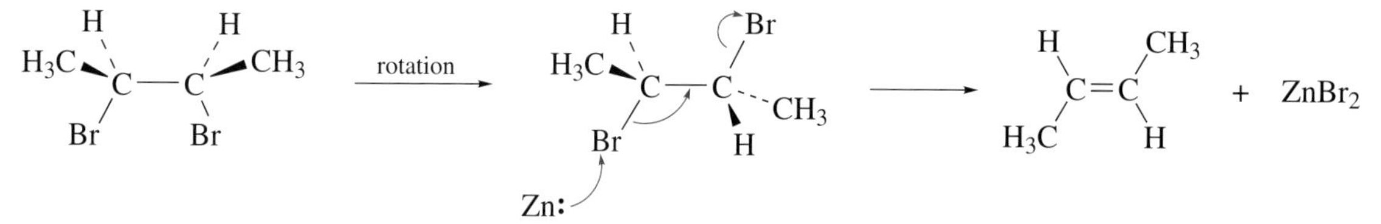

(b) In practice the major reaction product results from decarboxylative elimination, as shown in Reaction S.2. However, you may have suggested the elimination of HBr as a possibility, and on paper this is reasonable (and can lead to two different alkenes, one of which is capable of existing as *cis*- and *trans*-isomers).

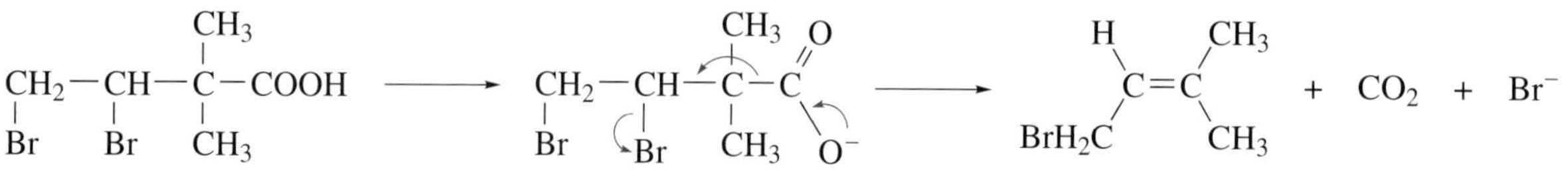

Alkynes (and the allene $CH_2=C=C(CH_3)_2$) would not be produced unless a stronger base was employed.

However, if you suggested the dehalogenation product, then you were wrong. Dehalogenation only occurs with iodide ion, or with a metal such as zinc.

ACKNOWLEDGEMENTS

Grateful acknowledgement is made to the following source for permission to reproduce material in this book:

Figure 2.1: Smith Image Collection, University of Pennsylvania.

Every effort has been made to trace all the copyright owners, but if any has been inadvertently overlooked, the publishers will be pleased to make the necessary arrangements at the first opportunity.

Teflon® is a registered trademark of E. I. du Pont de Nemours Inc.

Shape-selective Catalysis using Zeolites

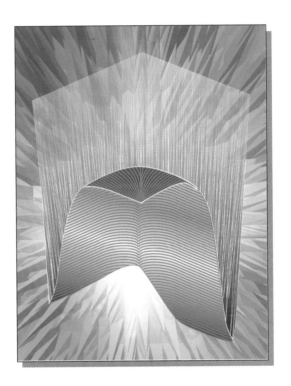

Craig Williams and Michael Gagan

INTRODUCTION

1

Zeolites are among the most widespread chemical materials in use today. They are of interest to geologists and mineralogists for their wide range, their occurrence and their mode of formation; and to chemists and crystallographers for their fascinating structures and their water-absorbing, ion-exchange and catalytic properties. They occur naturally as individual crystals and in huge deposits that can be mined; and they can be designed and created in the laboratory to fulfil specific chemical tasks.

The reason for the interest in zeolites lies in their unusual crystal structure at the molecular level. They are not solidly packed, like many crystalline materials, but they have continuous **pores** (**channels**) running through them, which intersect at **cavities** (**cages**) within the structure. These 'open spaces' can occupy up to 50% of the crystal volume. The diameters of these pores and cavities fall between 0.2 and 2.0 nm, leading to zeolites being classified as **microporous** materials.

The global output of natural zeolites in 1997 was around 3.6 million tonnes, 2.5 million tonnes of this coming from China. This usage was as lightweight building materials (2.4 Mt yr^{-1}), in agriculture, for animal feed supplements and soil improvers (0.8 Mt yr^{-1}), and for ion-exchange, adsorption, and catalyst applications in industry and the environment (0.4 Mt yr^{-1}). They are widely used to clean water in swimming pools and other waste waters, including the treatment of aqueous streams of nuclear waste, where they selectively take up radioactive strontium and caesium ions by ion exchange. The domestic market in the USA reaches about 35 000 tonnes annually, with more than half going for pet litter. However, for many years after their discovery, zeolites were mainly used in jewellery and as museum exhibits, owing to their exquisite beauty.

Most of the synthetic zeolites are used as detergent 'builders', to soften hard water by removing (mainly) the calcium ions that it contains. They were introduced when phosphates were banned from detergent formulations because of the pollution they caused. In terms of value, high-tech zeolite catalysts, often tailor-made and modified for specific applications, are at the top end of the market with a value of 500 million dollars a year in the USA alone. They are indispensable in the petroleum industry for producing sulfur- and nitrogen-free high-octane fuel (especially since the phasing out of lead additives); and in the plastics industry, especially for polyester fibres.

In this Case Study we shall look mainly at the structures and the chemical properties of some selected zeolites.

1.1 Natural zeolites

Zeolites are a mineral type first discovered in 1756 by Baron Axel von Cronstedt, a Swedish mineralogist. He noted that when he heated the mineral sample it hissed and bubbled, giving off steam. Consequently he named the minerals 'zeolites', derived from the Greek words 'zein' meaning to boil and 'lithos' meaning rock. Extensive mineralogical and geological studies since 1756 have given rise to the discovery of 56 naturally occurring zeolites, making them one of the largest known groups of minerals. The zeolite identified by von Cronstedt was later given the name stilbite. Figure 1.1 shows some of the many zeolite minerals that can be found in the UK, together with their dates of discovery. Worldwide, new zeolites are still being discovered, with goosecreekite (1980) and gobbinsite (1982) among the more curiously named recent finds.

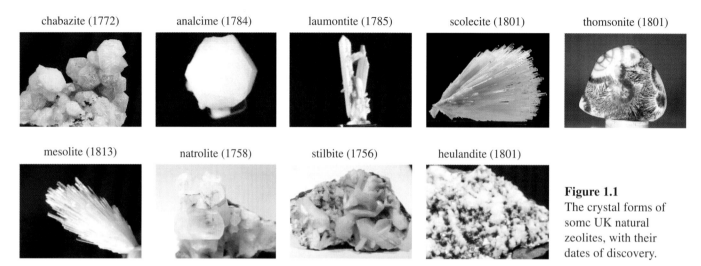

chabazite (1772) analcime (1784) laumontite (1785) scolecite (1801) thomsonite (1801)

mesolite (1813) natrolite (1758) stilbite (1756) heulandite (1801)

Figure 1.1
The crystal forms of some UK natural zeolites, with their dates of discovery.

Mineralogically, zeolites are hydrated aluminosilicate minerals, which originate during the formation of igneous, metamorphic, or sedimentary rocks, principally via hydrothermal processes. Hot circulating water leaches out mineral components deep within the Earth, particularly from basalt magmas. Zeolites form in these water-rich geological environments when conditions of low temperature ($<200\,°C$) and low pressure (4×10^8 Pa, equivalent to around 15 km depth of burial) are attained.

During the 1920s, scientists used the recently developed analytical technique of X-ray diffraction to determine the crystal structures of the various zeolites. These proved to be both interesting and unusual. All were shown to have three-dimensional open framework structures containing cavities and pores. Within these open spaces are located metal cations and the water molecules whose loss on heating first attracted the attention of von Cronstedt. The abundant, naturally occurring cations of sodium, potassium, magnesium, calcium and barium are the ions that occur most frequently, but depending on the zeolite almost all cations could be accommodated.

1.2 Synthetic zeolites

During the 1930s, scientists began to try to produce zeolites artificially in the laboratory. Early attempts concentrated on mimicking the conditions required to form natural zeolites. It was known that some natural zeolites had formed during the later stages of volcanic activity, when basaltic magmas had erupted and were cooling down. Consequently, early investigators tried treating powdered quartz with molten metal hydroxides at very high temperatures. Unfortunately, the majority of the final products thus formed were from another commonly occurring group of aluminosilicate minerals, the feldspars, rather than zeolites. However, in 1938,

Richard Barrer at the University of Aberdeen hit on the idea of using very reactive silica and alumina reagents, and lowering the reaction temperature to between 100 °C and 250 °C. In a typical reaction, he treated sodium hydroxide solution with aluminium foil to form a reactive sodium aluminate solution. He then added very fine amorphous silica, the most reactive source of silica available at that time, producing a grey gel. This was then stirred thoroughly at room temperature, and finally sealed in a suitable vessel and placed in an autoclave at around 150 °C for several days. Tiny crystals formed, which were filtered off and heated strongly to produce the zeolite. He found that by varying the amounts of reagents, the temperature, and the time of reaction, he could produce a whole series of different zeolites. His method of zeolite production in hot concentrated aqueous solutions with a high pH value, became known as the **hydrothermal method**, and it is still the principal method used today.

One effective method of pre-determining the type of zeolite structure that will be formed is the use of 'template molecules' of the appropriate size. Using bulky tetra-alkylammonium hydroxides rather than alkali metal hydroxides to increase the pH gives more open structures. For example, if a short-chain alkyl trimethylammonium surfactant (detergent), $C_6H_{13}N^+(CH_3)_3$, is used in the reaction mixture, this provides a template of a suitable size for the formation of the medium-pore zeolite ZSM-5 (Zeolite Synthesized by Mobil, number 5).

This technique has recently been developed for the synthetic formation of **mesoporous** molecular sieves (Section 6), which contain cavities with diameters between 2.0 and 20 nm, ten times larger than those in the microporous zeolites.

Professor Richard Maling Barrer, the 'Father of Zeolite Science'.

Richard Barrer, one of the pioneers of zeolite chemistry, was born on a New Zealand sheep farm, and studied at the same college as his famous scientific compatriot, the nuclear scientist Ernest Rutherford. He published his first paper on zeolites in 1938, the first of more than 400 papers and 21 patents in this area. In 1949 he succeeded in producing the first synthetic zeolite that had not, as yet, been discovered in nature. This was zeolite MAP, which is now widely used in washing powders to help to control the formation of scum during an automatic wash cycle, by adsorbing the calcium ions found in hard water. Barrer went on to produce well over 30 other zeolites, many by using the 'template synthesis', which he developed. He was a very active man and won the British Universities cross-country championship while studying for his PhD at Cambridge. He almost made the New Zealand team, at 10 000 metres, for the Berlin Olympics in 1936, and played tennis well into his 80s. He settled in the UK in 1937, and stayed for nearly 60 years, becoming Head of Department at Imperial College, London. He was nominated for a Nobel Prize in 1996, but died before the awards were announced.

STRUCTURE, PROPERTIES AND CLASSIFICATION OF ZEOLITES

2

2.1 Basic structures

A zeolite's three-dimensional structure is built up by linking together silicate $[SiO_4]^{4-}$ and aluminate $[AlO_4]^{5-}$ tetrahedra (Figure 2.1a), by the sharing of oxygen atoms at all four corners.

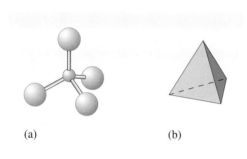

(a)   (b)

Figure 2.1
Representations of the TO_4 unit
(T = Si or Al) as (a) a tetrahedral bond
structure and (b) a solid tetrahedron.

The tetrahedron on its own does not give much of an idea of the three-dimensional structures that can be constructed by connecting the tetrahedra together. You should already be familiar with the quartz (silica) structure of SiO_2, which consists of tetrahedra of $[SiO_4]^{4-}$ linked to each other through all four oxygen atoms (Figure 2.2a). An alternative way of looking at this structure is as a giant molecule in which every silicon atom is joined through an oxygen atom (the 'oxygen bridge') to four other silicon atoms (Figure 2.2b).

Zeolite frameworks are constructed on a similar pattern. Their structures become clearer to visualize when tetrahedra are assembled into the cage structure of the mineral sodalite. This still looks complex when both the silicon (aluminium) *and* the oxygen atoms are shown. Structural representations are considerably simplified by omitting the atoms altogether, as in Figure 2.3. The silicon (aluminium) atoms are located at every *corner* of the structure, and the oxygen bridges between them are shown as straight lines. Remember that this is not an entirely accurate representation, as the Si—O—Si angle will never be 180°.

The cavity within the sodalite structure has two types of entry point: the square port and the hexagonal port. The mineral sodalite has these cages packed directly together by sharing the silicon atoms of the square ports, to give an extended structure with the cavities directly connected to each other (Figure 2.4a overleaf). This is the optimum space-filling arrangement, and leads to a structure that also has a sodalite cage at its heart. In zeolite structures, the silicon/aluminium tetrahedra are connected in a myriad different ways, but the sodalite structure is often a building block. In the structure termed zeolite A, for example, each square port of silicon/ oxygen atoms in the sodalite cage is connected to another square port through oxygen bridges (Figure 2.4b), whereas in zeolites X and Y sodalite cages are linked by oxygen bridges between the silicon/aluminium atoms of the hexagonal ports (Figure 2.4c).

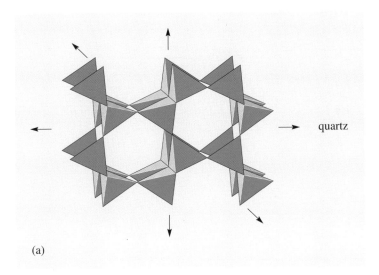

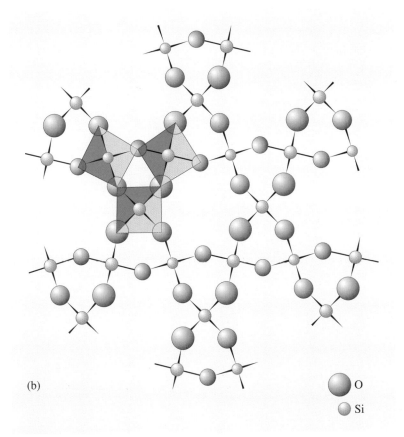

O O

Si

Figure 2.2 The structure of quartz, represented as (a) linked tetrahedra and (b) an extended molecule.

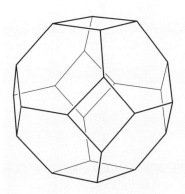

Figure 2.3 A representation of the sodalite cage as a line structure.

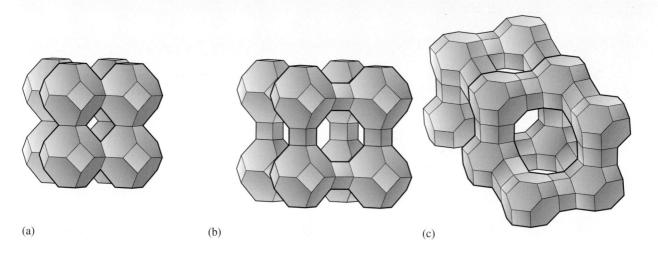

Figure 2.4 The construction of zeolite frameworks: (a) directly connected sodalite cages in the mineral sodalite; (b) zeolite A (LTA), with oxygen bridges between square ports; (c) zeolite X (or Y) (FAU), with oxygen bridges between hexagonal ports.

The size of the pore is determined by the number of tetrahedra that make up the cross-section of the pore walls within a zeolite framework. For example, if four tetrahedra are linked together in a ring, a very small pore about 0.26 nm in diameter is produced (Figure 2.5a). If the cross-section is made up of six tetrahedra linked together in a ring then the diameter increases to about 0.33 nm (Figure 2.5b).

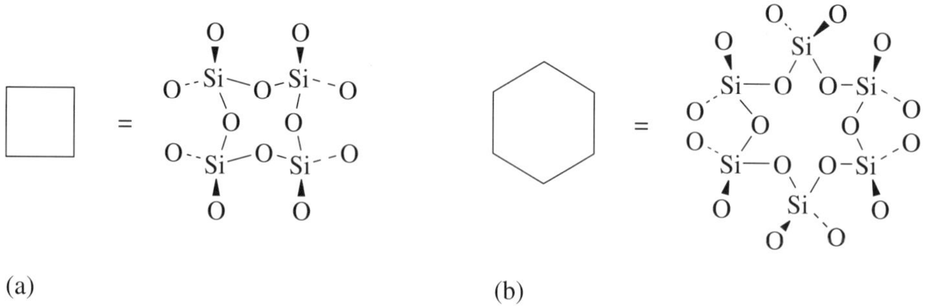

Figure 2.5
Pore cross-sections constructed from (a) four silicate tetrahedra and (b) six silicate tetrahedra.

2.2 Zeolite properties

It is this possibility for variation of the pore size that leads to an important property of zeolites, that of **molecular sieving**. A zeolite is able to accommodate or reject molecules based on their size. Using suitable drying methods, the water within a zeolite framework can be removed; the resulting dehydrated zeolite has space to accommodate other small molecules. So, if we have a mixture containing some molecules of suitable size and shape to enter the pores of the zeolite, and other molecules that are not able to do so, an effective separation can be carried out. A mixture of straight-chain and branched-chain hydrocarbons is one such example.

This property also makes particular zeolites very effective **desiccants**. If a sample of an organic compound, with molecules that are too large to enter the pores of a zeolite, contains molecules of water, which *can* enter these pores, the water is effectively 'scavenged' from the sample. Once they enter the hydrophilic interior of the zeolite the water molecules are unlikely to migrate out into the hydrophobic organic environment.

This can be taken further when two molecular reactants are small enough to enter the pores of a zeolite. If they undergo a reaction, promoted by the acidic sites (see below) within a cavity, when they reach it together, the final products of that reaction are dictated by the size of the zeolite cavity and the diameter of the pores. Even if more than one product is possible, if only one of them is of a suitable size or shape to escape from the interior of the zeolite, that will be the preferred product. This has enormous implications for reactions using zeolites as catalysts, and is known as **shape-selective catalysis**.

It should be noted that the silicate tetrahedra have a formal charge of 4−, but the aluminate tetrahedra have a formal charge of 5−. When a silicate tetrahedron is linked to four other silicate tetrahedra, as in quartz, this produces a neutral silicate framework. However, the insertion of aluminate tetrahedra, in place of silicate, introduces a charge imbalance into the structure. There will be one negative charge on the framework for every replacement of a silicate $[SiO_4^{4-}]$ tetrahedron by an aluminate $[AlO_4^{5-}]$ tetrahedron, because the aluminium atom has one less proton than the silicon atom, so one extra electron will have to be present to complete the covalently bonded structure (Figure 2.6).

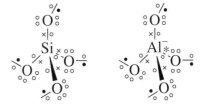

Figure 2.6
Lewis structures of silicate and aluminate tetrahedra in a zeolite framework, showing the presence of an extra electron (*) in the latter.

This negative charge on the framework must be neutralized by cations present in both the cavities and the pores. The presence of cations counterbalancing the negative charge on the framework leads to large electrostatic forces within the zeolite structure. It is these forces that encourage water to reside within the framework. The presence of the labile water and the cations creates the **ion exchange** properties of zeolites.

The sodium form of zeolite A, for example, has sodium ions attached to the inside of its pores and cavities, balancing the negative charges arising from the substitution of silicon by aluminium. Treatment of the zeolite with concentrated solutions of calcium chloride replaces the lining of Na^+ ions by one of Ca^{2+} ions. This is a reversible process, for if the resulting calcium form of the zeolite is treated with concentrated aqueous sodium chloride, sodium ions replace the calcium ions now adsorbed to the pore surfaces of the crystal structure. Water softeners work this way, and zeolite A is a common component of modern detergents (Figure 2.7).

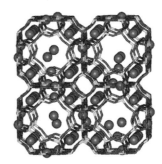

Figure 2.7
Computer representation of sodium ions adsorbed in the pores of zeolite A.

There is also another feature of zeolites that encourages other molecules to become adsorbed within the zeolite framework. In some zeolites not all the tetrahedra are fully linked together, and where some tetrahedra have only three or two linkages to others, a defect site is said to exist. At these defect sites the uncoordinated linkages are terminated by OH groups (silanol groups), which account for the so-called 'acidic sites' within a zeolite framework. The electron-withdrawing effect of the oxygen atom attached to the terminal hydrogen atom will be enhanced by a similar electron-withdrawing effect from the other three oxygen atoms attached to silicon. The O—H bond is therefore weakened, making the hydrogen acidic.

2.3 Zeolites as catalysts

The most remarkable property of zeolites is their activity as heterogeneous catalysts, especially in the cracking process used in the refining of petroleum, and in alkylation and isomerization reactions. Cracking is the reaction in which long-chain hydrocarbons are broken down to smaller molecules, at high temperatures and pressures over a catalyst. These smaller hydrocarbons are more useful as fuel in the internal combustion engine. It is claimed that every gallon of gasoline consumed in the USA has passed through a process that uses zeolites. Acidic clays and amorphous silica/alumina preparations have been used for many years for cracking, and it is known that the catalytic activity relates to tetrahedral aluminate units with associated H^+ ions, the acidic sites. However, until the open framework zeolites were available the number of active sites on the catalysts was limited.

Because zeolites can also be manufactured with various proportions of aluminate, a catalyst can be tailored to meet the exact requirement of the process. It is calculated that the medium-pore zeolite ZSM-5 (🖳), operating at 454 °C and 100 torr (1.3×10^4 Pa) pressure of hexane, can crack more than 37 molecules per active site per minute. At 538 °C the turnover rises to over 300 molecules per minute per active site. Other catalytic processes – toluene disproportionation, xylene isomerization, and methanol conversion (see later) – operate even faster, with hexane isomerization showing a turnover of as much as 4×10^7 per minute per active site. This indicates that rates of catalytic reactions with zeolites equal or exceed rates for enzyme catalysis.

Many of the reactions catalysed involve the transfer of protons to and from the zeolite (Figure 2.8). As this is the characteristic function of an acid, the H-forms of the zeolite (those in which the adsorbed ions are H^+) are sometimes referred to as 'solid acid' catalysts. Synthetic zeolites are often prepared in concentrated solutions of sodium hydroxide, which leaves Na^+ as the counter ion in the zeolite pores, the Na-zeolite. The usual way to prepare H-zeolites is to exchange these metal ions for ammonium ions (NH_4^+), and then heat the ammonium form of the zeolite to about 500 °C. Ammonia (NH_3) is then driven off, leaving H^+ ions to balance the negative charges on the silicate structure. This is the reverse of the reaction shown in Figure 2.8, when the base B is NH_3.

The solid acids are the most used of all heterogeneous catalysts. As they have to replace traditional inexpensive mineral acids, they have to be cheap. However, costs are offset by the problems encountered when using highly corrosive and waste-generating acids, which have to be neutralized before disposal.

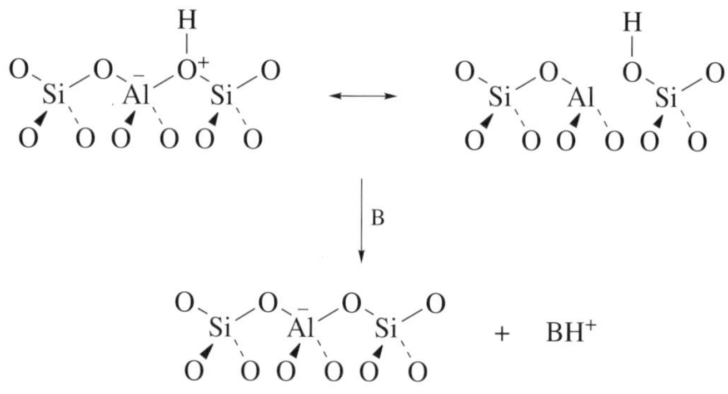

Figure 2.8
Proton association and transfer in zeolites (B = base, removing a proton from an acidic site). 🖳

Both proton-donating groups and electron-accepting centres (Lewis acids) are present in zeolites. The silanol form of the aluminosilicate (Figure 2.8, top right) shows aluminium in its normal trivalent state, in which the aluminium is electron-deficient, being two electrons short of the stable octet. As a result, there are centres in the aluminosilicate structure that act as Lewis acids, in the same way as aluminium chloride or boron trifluoride. Thus there are sites with different acid strength within the zeolite, and its ability to catalyse a particular reaction may depend on the number and distribution of sites having the necessary acid strength.

2.4 Zeolite classification

In Table 2.1 a few typical, and important, zeolites are listed together with their pore dimensions. The various zeolite structures are now given a three-letter code. Apart from mordenite, which has useful pores in two dimensions only, all the other zeolites in Table 2.1 have pores in three dimensions.

Table 2.1 Key properties of some common useful zeolites (🖳).

Molecular sieve type	Structure type code	Largest pore dimensions/nm	Number of Si/Al tetrahedra in pore cross-section
Small pore			
zeolite A	LTA	0.41 (circular)	8
erionite	ERI	0.36×0.51	8
Medium pore			
ZSM-5	MFI	0.53×0.56	10
ZSM-11	MEL	0.53×0.54	10
Large pore			
zeolite X and Y (faujasite)	FAU	0.74 (circular)	12
mordenite	MOR	0.65×0.70	12

The structure type code was introduced by IUPAC (International Union of Pure and Applied Chemistry) to try to simplify the notation of zeolite structures. In the past several researchers had produced the same basic zeolite by different laboratory methods. The zeolite thus obtained was named by that researcher and resulted in a system where one zeolite structure was known by up to 20 different trade names! This led to a great deal of confusion and so the three-letter code was introduced. So for example zeolite A (produced using sodium hydroxide, alumina and silica) and zeolite ZK4 (produced using tetramethylammonium hydroxide, alumina and silica) have the same structure, now denoted LTA (🖳).

Many zeolites have more than one pore structure, and Table 2.1 lists the dimensions of the largest pores in nanometres. These pores can be circular, as in zeolite A (LTA) with a diameter of 0.41 nm, whereas others are elliptical, as in erionite (ERI; 🖳) with dimensions 0.36×0.51 nm. As well as having different size circular or elliptical pores, zeolites may also have pores that run in just one dimension, or in two or three dimensions. We can see that with this wide range of pore sizes and orientations, the separation of various substances with different molecular sizes can be achieved.

An additional consideration is the degree of silicon substitution by aluminium within a zeolite structure. This will alter the charge, usually termed the polarity, on the zeolite's framework. The amount of aluminium substitution is usually expressed as the ratio of silica (SiO_2) to alumina (Al_2O_3). Low SiO_2/Al_2O_3 ratios, indicating high levels of aluminium substitution, tend to make a zeolite hydrophilic (water liking), so when exposed to a mixture of, say, water and alcohol, the zeolite will take up the water in preference to the alcohol. High SiO_2/Al_2O_3 ratios make the zeolite hydrophobic (water disliking), so when the zeolite is exposed to a mixture of water and alcohol the zeolite will take up the alcohol in preference to the water.

The hydrophilic nature of the zeolite increases with the substitution of aluminium atoms for silicon atoms because substitution increases the negative charge on the framework. When this happens, positively charged metal ions will be adsorbed in the pores, and so both the negative framework and the positive cations will attract the polar water molecules.

Figure 2.9 shows the various silica/alumina ratios for some common zeolites plotted against the aluminium content. They range from zeolite A with a ratio close to 1 : 1, through zeolite X (from 2 : 1 to 3 : 1), zeolite Y (from 3 : 1 to 6 : 1), and mordenite (10 : 1), to ZSM-5 with ratios from 10 : 1 to infinity (indicating all silica and no alumina, a structure known as a silicalite). This series also ranks these zeolites in the order of their hydrophilic nature, with the most hydrophilic (zeolite A) first.

This variation in polarity can also have an influence on the way in which organic molecules and their reaction products are attracted to the inner surfaces of the pores and cavities of zeolites (**adsorption**) and released from them (**desorption**).

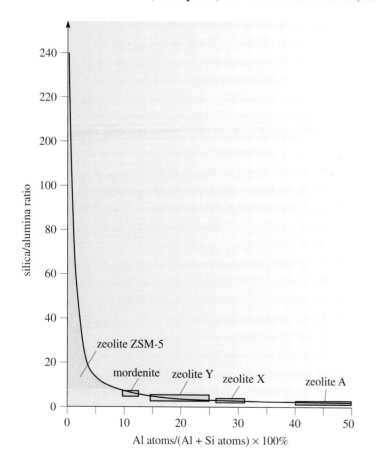

Figure 2.9

Plot of silica/alumina ratio *versus* aluminium content for selected zeolites

However, the influence of polarity in molecular sorption is small compared with the influence of pore size.

Zeolites offer a richly diverse range of pore size, pore connectivity, framework polarity and acidic site concentration. In addition, the external morphology of synthetic zeolite crystals may be modified through the use of organic co-solvents during crystallization. The crystal morphology of a zeolite will influence its final use. For example, a crystal with a polygonal shape is appropriate to provide the small spheres used in fluidized-bed reactors, where the crystals are constantly tumbling over each other. However in a fixed-bed reactor where the crystals do not move, a cubic-type morphology is preferred as it allows regular packing and a freer movement of the reactants through the fixed bed.

2.5 Small-pore zeolites

The small-pore zeolites (Table 2.1), with pore cross-sections constructed from eight alternating silicon and oxygen atoms (Figure 2.5a), typically accommodate linear molecules, such as straight-chain saturated hydrocarbons, or primary alcohols and amines. They do not allow access by branched isomers. Zeolite A is a typical small-pore zeolite with large cavities (Figure 2.10). If the silica/alumina ratio of zeolite A is held at 2 : 1, in its sodium form the cross-sectional diameter is 0.41 nm, compared with a cavity diameter of 1.14 nm. This is why only linear organic molecules may *enter* the pores of zeolite A. Once *inside* the structure, molecules may react within the cavities to produce much larger intermediates/transition states. However, the final products must be small enough to allow them to exit from the zeolite.

The pore system of zeolite A can be widened by removing some of the sodium ions normally present and replacing them with calcium ions. The result of this ion exchange is to increase the effective pore size to 0.5 nm; conversely, if the sodium is replaced by potassium this is reduced to 0.3 nm, allowing only water molecules into the cavities in the system. This alteration of the dimensions of pore systems by cations is termed the 'sentinel effect'. The simplest explanation of why the replacement of sodium ions changes the diameter of the pores in zeolite A is that K^+ ions are larger than Na^+ ions. Also, although the sizes of Na^+ and Ca^{2+} are virtually identical, only half as many Ca^{2+} ions will be needed to balance the negative charge on the zeolite framework when they replace Na^+ ions. So the space down the centre of the pore when these exchanges are made will change accordingly.

Figure 2.10

The structure of zeolite A (LTA): (a) line drawing; (b) eight-membered ring cross-section viewed along the direction of the arrow in (a); (c) schematic representation of the pore system.

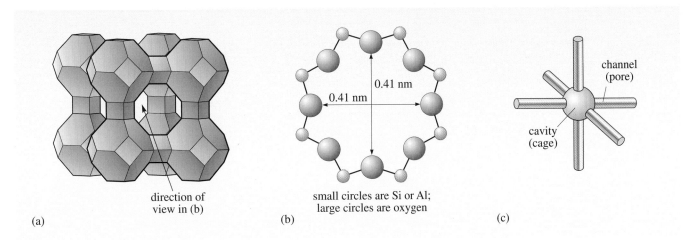

direction of view in (b)

0.41 nm   0.41 nm

small circles are Si or Al; large circles are oxygen

channel (pore)

cavity (cage)

(a)   (b)   (c)

2.6 Medium-pore zeolites

The cross-sectional structure of the medium-pore zeolites (Table 2.1) is a 10-membered ring of alternating silicon and oxygen atoms. The pore systems tend to have uniform dimensions, without large cavities. The intermediate size of their pores enables a great deal of discrimination in molecular transport and chemical reaction. A typical zeolite of this type is ZSM-5 (Figure 2.11), which has two perpendicularly intersecting 10-membered-ring pore systems that form a three-dimensional network. One type of pore is straight and nearly circular (0.53×0.56 nm), and the other is sinusoidal and more elliptical (0.51×0.56 nm). At the intersections the spherical diameter of the free volume is around 0.9 nm, and this space has great significance in catalysis. ZSM-5 can be synthesized with silica/alumina ratios from around 5 to 8 000 (Figure 2.9), which enables the zeolite to show progressively more hydrophobic character as the aluminium content decreases.

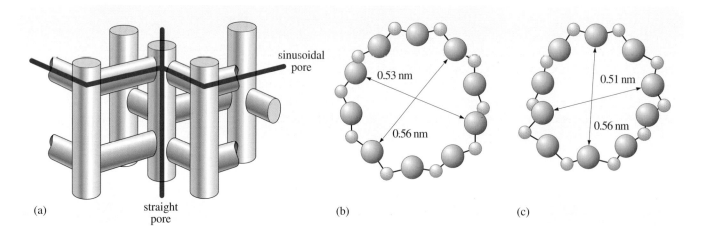

(a) straight pore sinusoidal pore   (b)   (c)

Figure 2.11 The structure of ZSM-5 (MFI): (a) schematic representation of the intersecting straight and sinusoidal pores; (b) the 10-membered ring from a straight pore; (c) the 10-membered ring from a sinusoidal pore.

2.7 Large-pore zeolites

The presence of pore structures with a cross-section of 12 alternating silicon and oxygen atoms is characteristic of large-pore zeolites (Table 2.1). Mordenite is such a zeolite (Figure 2.12), with a two-dimensional pore system containing one set of large pores (0.65 × 0.7 nm) cross-connected with a set of smaller pores with eight-membered-ring cross-sections, creating 'side pockets'. Unfortunately, mordenite suffers from extensive stacking faults, which block the free movement of molecules within the structure and make it less useful as a catalyst.

The other well-known large-pore zeolite is faujasite (Figure 2.13 overleaf). The synthetic versions are known as zeolite X and Y, with zeolite Y having a higher silica/alumina ratio than zeolite X (Figure 2.9). The faujasite pore system is very open and three dimensional. It is constructed of sodalite cages connected by oxygen bridges through the six-membered-ring faces. This gives an array of nearly spherical cavities (1.14 nm diameter) each of which is accessed via four tetrahedrally orientated 0.74 nm ports. The overall pore structure, which is comparable to the carbon–carbon bonding structure in diamond, is shown in Figure 2.13d.

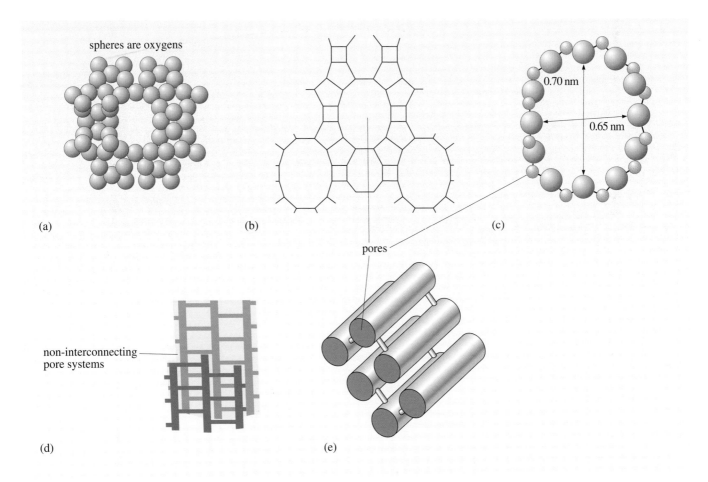

Figure 2.12 The structure of mordenite (MOR): (a) cross-section of a large pore, showing the oxygen atoms; (b) projection of the framework; (c) 12-membered ring (from the same view as the cross-section); (d) schematic representation of the two-dimensional pore system; (e) simplified view of the large-pore system as an assembly of parallel tubes.

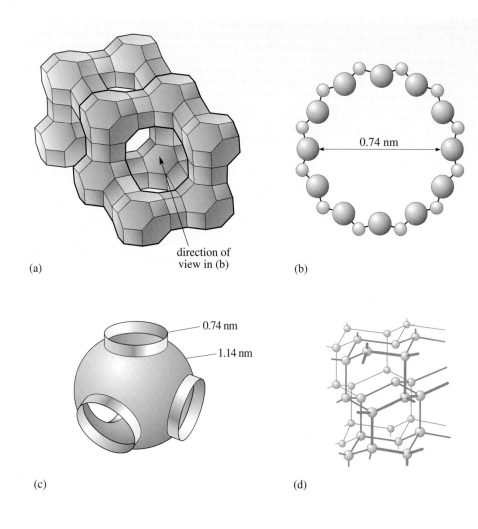

(a)

direction of view in (b)

(b)

0.74 nm

(c)

0.74 nm
1.14 nm

(d)

Figure 2.13
The structure of faujasite (FAU):
(a) line drawing; (b) 12-membered
ring viewed along the direction
shown in (a); (c) simplified diagram
of the cavity with its four 12-ring
ports; (d) the diamond-like pore
system.

COMPUTER ACTIVITY 2.1
Zeolite structures on WebLab ViewerLite

The CD-ROM associated with this book contains several structures of zeolites
that can be viewed using WebLab ViewerLite: erionite (ERI), faujasite (FAU),
zeolite A (LTA), mordenite (MOR), ZSM-5 (MFI) and ZSM-11 (MEL).

They are saved as line drawings of the unit cell, but in order to get a better view
of the structure you must first obtain the structure as a continuous array.

- Start WebLab ViewerLite. Open the file ERI.msv from the Figures/Zeolites
  folder on the CD-ROM.

- From the Tools menu, select Crystal Cell, and on the Preferences tab,
  increase the A, B, and C numbers in the View Range boxes. Change the
  values to A = 2, B = 2, and C = 2. This generates a view of eight unit cells.

- Click the Fit to Screen button on the horizontal toolbar. The new structure
  should now be centred in the screen with the atoms visible.

- From the View menu, choose Options, and on the Graphics tab check that
  Quality is set to Medium and the 'Fast render on move' option is ticked.
  These line drawings may easily be rotated and translated in this form.

However, it is more instructive, and gives a more realistic view of the pores and their comparative dimensions, if you convert the line drawings into space-filling models.

- From the View icon on the toolbar, click Display Style and then select CPK in the dialogue box.

You can also view the van der Waals radii.

- From the Tools menu, select Surfaces, then Add.
- From the View menu, select Display style, choose the Surfaces tab, tick VDW and transparent, and then click OK.

Once you have the extended array form, it is interesting to change the display from Orthographic to Perspective (under View|Options|Graphics). This gives the impression of the array receding from you, and is particularly effective with large arrays.

Now measure some of the internal pore distances in erionite.

- Return the display style to Line. Rotate the model until you are looking directly down a pore; you may prefer to revert to the Orthographic view for clarity.
- Click on the Select tool from the vertical toolbar, then on an oxygen at one side of the pore. With the Shift key depressed, click on an oxygen across the pore diameter.
- From the Tools menu, select Monitors and from the submenu choose Distance.

The diameter of the pore is given in Ångstroms (divide by 10 for nanometres). Rotate the model and find the diameters of several other pores to explore the range of pore sizes in erionite. Note that WebLab ViewerLite measures *interatomic* distances. These will be larger than the pore diameters quoted in the text, which have taken the van der Waals radii into account. If you have time, try out some of the other zeolite structures from the CD-ROM.

One final idea you may wish to pursue is to view the zeolites in stereoscopic form.

- From the View menu, select Options and in the dialogue box select the Stereo tab, and also select Split Screen Stereo from the View menu. You are presented with two side-by-side stereoviews of the zeolite model.
- Use the 'Fit to Screen' icon to get two suitably sized, centred images
- Adjust the window sizes so you can take both images into view as you come close to the monitor screen.
- Try both the Crossed eyes and the Relaxed eyes options in the Stereo dialogue box to find your preferred way of viewing. You may need to use a card or something similar (a floppy disk?) to separate the field of view for each eye, before you can achieve a stereo view.field of view for each eye, before you can achieve a stereo view.

SHAPE SELECTIVITY

3

A major characteristic of zeolites is shape selectivity, the ability of different frameworks to accommodate only molecules of a particular shape and size within the pores of their structure, with a high degree of specificity. The medium-pore zeolites demonstrate this phenomenon more than any of the other zeolites mentioned. This Section describes three ways in which zeolites can bring about this shape selectivity.

3.1 Mass-transport discrimination

The ability of a molecule to pass through the pores of a zeolite is called diffusion. The degree to which each type of molecule possesses this ability is termed its diffusivity. Mass-transport discrimination arises when there are large differences in diffusivities among the various classes of molecules with respect to pore size. This process can result in complete exclusion of the molecule from the zeolite pores. Figure 3.1 shows how linear hydrocarbons can enter the pores of the medium-pore zeolite ZSM-5, but branched-chain hydrocarbons are excluded. Molecules entering the pores of a zeolite may not survive their passage over the highly catalytic surface, and Figure 3.1 indicates that some breakdown of the hydrocarbon into smaller molecules (termed *cracking*) can occur.

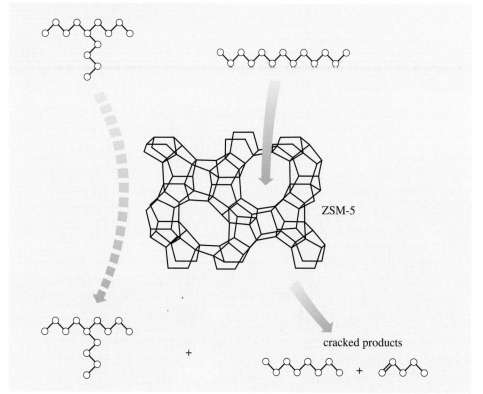

Figure 3.1
When a mixture of linear and branched saturated C_{13} hydrocarbons is passed over a medium-pore zeolite, such as ZSM-5, only the linear molecules enter the pores, and may then undergo breakdown.

Mass-transport discrimination is not limited to total exclusion, but can also occur when several species are able to enter the pore system. If the diffusivity of each species in a mixture is sufficiently varied, then the different molecules will diffuse through the zeolite at different rates, and can be separated. This is shown in Figure 3.2.

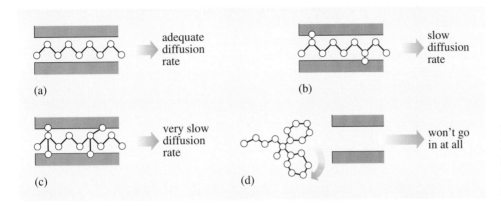

(a)

(b)

(c)

(d)

Figure 3.2
Relationship between molecular size and shape and diffusivity in the pores of a medium-pore zeolite, such as ZSM-5.

This 'molecular sieving' effect when alternative reaction products may be formed can also dictate which reactions are able to take place within the pores of a zeolite. This is shown diagrammatically in Figure 3.3 for isomerization *versus* cracking of hydrocarbons in zeolite A, and for preferential formation of 1,4-dimethylbenzene (*para*-xylene) in ZSM-5.

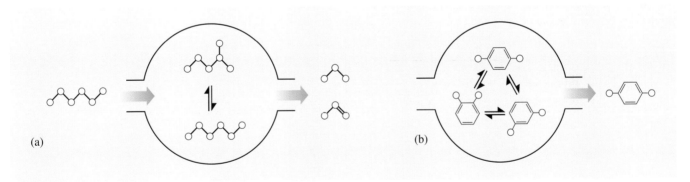

(a)

(b)

Figure 3.3 Product selectivity for (a) cracked hydrocarbons in a small-pore zeolite (zeolite A), and (b) *para*-xylene in a medium-pore zeolite (ZSM-5)

Isomerization of a straight-chain hexane to a branched isomer can occur in the 1.14 nm cavity of zeolite A, but the branched hydrocarbon product cannot leave through the 0.4 nm pore (Figure 3.3a). It is sometimes possible for these bulkier molecules to be removed via further reaction to yield sterically less demanding isomers, but when this is not possible the pore system eventually becomes blocked.

Restricting a reaction to produce a specific product, when more than one product is possible, can also occur because of the differences in the relative diffusivities of the various products. In Figure 3.3b, *para*-xylene can emerge from the pore system much more rapidly than either the *meta* or the *ortho* isomer. A representation of *para*-xylene in a zeolite pore is shown in Figure 3.4.

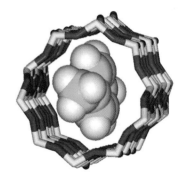

Figure 3.4
Computer representation of *para*-xylene molecule in ZSM-5.

Even if *ortho-* and *meta-*xylene form readily within the cavity, their inability to escape means that they will remain and isomerize to the more mobile *para-*isomer. This is commercially valuable because the world demand for *para-*xylene is constantly increasing, owing to its use as an intermediate in the production of polyester (Terylene).

Figure 3.5 shows the correlation between the pore sizes of various zeolites and some molecular diameters; the yellow areas indicate the range of diameters of the zeolite pores. We see that zeolite A has a single type of pore, with an internal diameter of 0.4 nm, whereas mordenite has pores of two sizes, one 0.30×0.57 nm, and the other 0.65×0.70 nm. The yellow area for ZSM-5 shows the size-overlap of its two distinct pore systems. Molecules with diameters falling within the yellow zones (or in the pale pink zones below them) can be accommodated in the pores of the zeolite. However, if the molecular diameter falls within the darker pink zones, those molecules will be excluded.

1,2-dimethylbenzene
*ortho-*xylene

1,3-dimethylbenzene
*meta-*xylene

1,4-dimethylbenzene
*para-*xylene

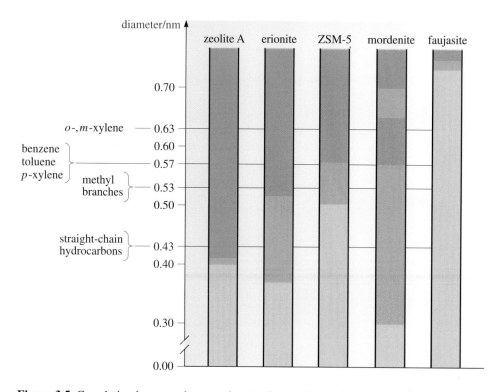

Figure 3.5 Correlation between the pore sizes (yellow regions) of various zeolites and some molecular diameters.

Figure 3.5 allows us to predict, for example, that hydrocarbons with methyl branches (molecular diameter 0.53 nm) would have too great a diameter to pass through the pores of erionite (internal dimensions 0.36×0.51 nm), and that faujasite (internal diameter 0.74 nm) could easily accommodate *ortho-* and *meta-*xylene (molecular diameter 0.63 nm), as well as the slimmer *para-*xylene.

3.2 Transition-state selectivity

There are a number of instances where it would be expected that both reactant and product molecules should be able to diffuse rapidly in and out of a zeolite, but nevertheless some product selectivity is observed. It is well known that the activated complex (or transition state) in a bimolecular reaction is usually larger than either the reagent or the product molecules. So it has been proposed that selectivity results from the inability of the reactants to attain a transition state within the confines of the zeolite structure. An interesting example is the reaction of 1-ethyl-2-methylbenzene over mordenite at 204–315 °C, in which two molecules combine within the mordenite cavity. 1-Ethyl-2-methylbenzene readily isomerizes to 1-ethyl-3-methylbenzene under these conditions, but only the former undergoes further reaction to give a trialkylbenzene. There are two possible transition states, producing two possible products, but only one of them is formed. This implies that the zeolite is restricting which transition state can form. This reaction is shown in Figure 3.6. It can sometimes be difficult to decide whether selectivity is due to product shape or to transition-state size and shape.

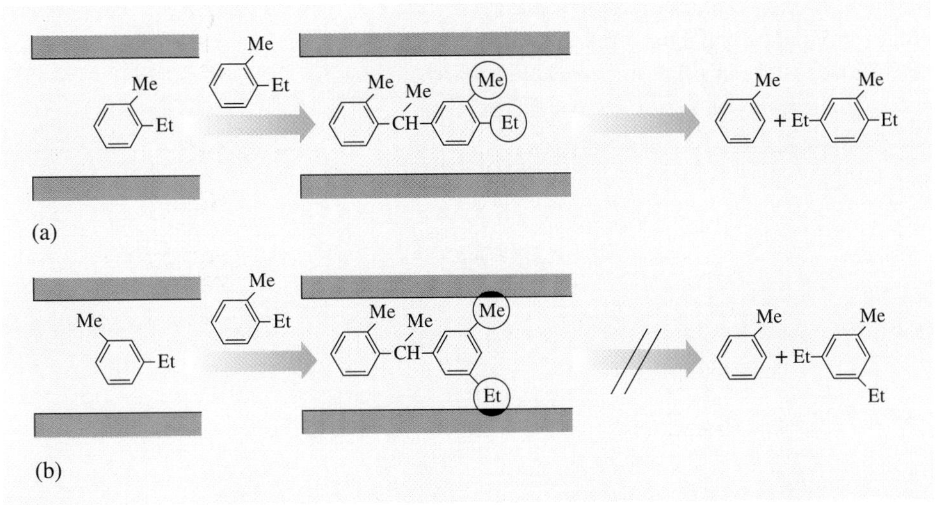

Figure 3.6
Restrictive transition-state selectivity in the reaction of 1-ethyl-2-methylbenzene over mordenite: (a) the transition state can be formed; (b) the isomerization of 1-ethyl-2-methylbenzene to 1-ethyl-3-methylbenzene means that the geometric and volume requirements of the transition state are incompatible with the available pore space. Note that, in this diagram, Me = CH_3 (methyl) and Et = C_2H_5 (ethyl)

3.3 Molecular traffic control

Several zeolites have intersecting pores that differ in size (e.g. ZSM-5 and mordenite). In this situation the small pores will accept only small molecules whereas the large pores will take both small and large molecules. This creates a different type of shape selectivity, which has been described as 'molecular traffic control'. It has been used to explain the absence of counter-diffusion in ZSM-5 during the conversion of methanol into gasoline. The methanol molecules enter ZSM-5 via the smaller sinusoidal pores, and the larger hydrocarbon products leave via the larger straight pores (see Figure 2.11).

APPLICATIONS OF SHAPE SELECTIVITY

4

There are now several industrial processes that use the shape selectivity afforded by zeolite catalysts. These include aromatic substitution, aromatic alkylation and alkylbenzene isomerization, and methanol to gasoline conversion reactions. A number of reactions deserve special mention.

4.1 *Para* selective alkylation of aromatic hydrocarbons

The substitution of a hydrocarbon side chain into a benzene ring is called *aromatic alkylation*. Using zeolite ZSM-5, modified by the inclusion of phosphate ions ($PO_4{}^{3-}$), at 500 °C, *para*-xylene has been synthesized selectively with a purity of up to 97%. A typical aromatic alkylation of monoalkylbenzenes uses a toxic and hazardous catalyst, such as $AlCl_3$ or $FeCl_3$, and results in predominantly *ortho-para* substitution. Not only that, but rapid secondary alkylation reactions usually generate a mixture of products with two, three or more alkyl substituents (Figure 4.1).

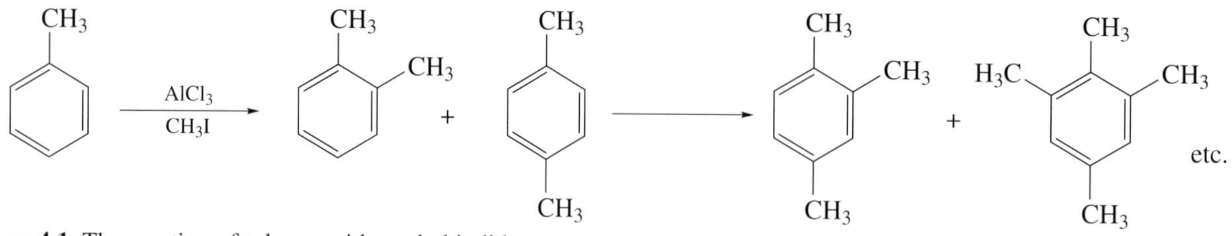

Figure 4.1 The reaction of toluene with methyl iodide.

This reaction is also rather messy, generating a large amount of aqueous waste products containing aluminium, giving a low overall yield, and requiring organic solvents. Zeolites on the other hand are very 'green' *, not only giving good yields of a single *para*-disubstituted product (Figure 3.3b), but also allowing the zeolite to be regenerated for reuse.

4.2 Selective xylene isomerization

As *para*-xylene is the most valuable xylene for use as a polymer intermediate, it is useful that *ortho*- and *meta*-xylene can be isomerized using zeolite catalysts. Two routes are possible: *disproportionation*, that is swapping methyl groups between xylene molecules; and direct *isomerization*, involving sequential shifts of methyl groups around the benzene ring.

---

* 'Green chemistry' refers to a recent move to make industrial chemistry environmentally friendly by, for instance, decreasing the amounts of toxic by-products and waste.

Using the large-pore zeolites X and Y (faujasite), disproportionation of xylenes is the principal route, at *fifty times the rate* of direct isomerization. It takes place via intermediates in which two benzene rings are linked together through an alkyl group, a *bimolecular* reaction (*cf.* Figure 3.6). This is easily possible in the larger cavities of faujasite.

With medium-pore zeolites (e.g. ZSM-5) very little disproportionation occurs, because the smaller cavities of ZSM-5 cannot easily accommodate this bulky intermediate. Much more of the product is formed through the isomerization reaction proceeding via a mechanism in which only a single molecule of xylene is involved. In this *unimolecular* mechanism, xylene undergoes a 1,2-methyl shift (Figure 4.2). The benzene ring is protonated by an acidic site in the zeolite, at the carbon atom carrying the methyl substituent, and the methyl group (with its pair of electrons) migrates to an adjacent carbon atom of the ring. Loss of hydrogen ion back to the zeolite generates the isomer.

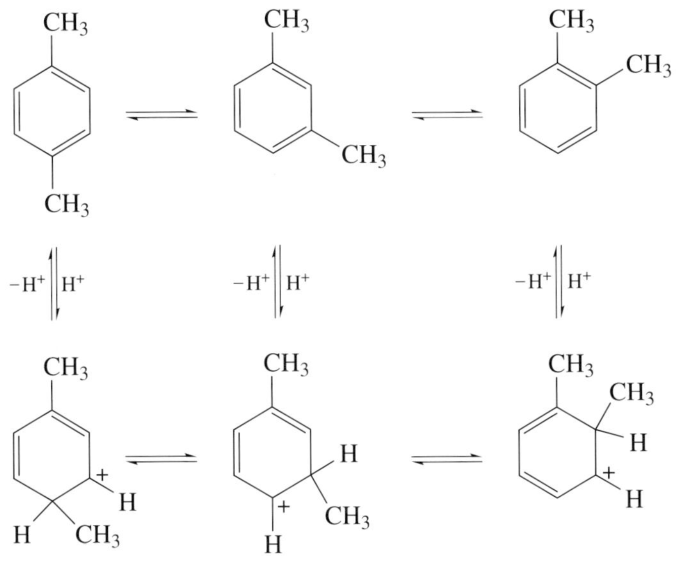

Figure 4.2
Xylene isomerization over ZSM-5 featuring sequential 1,2-methyl shifts.

4.3 Some other selective alkylation reactions of aromatic compounds

The zeolite-catalysed alkylation of phenol or anisole (methoxybenzene) with methanol is a complex sequence of reactions taking place within ZSM-5 and ZSM-11 at 200–300 °C. As shown in Figure 4.3 (overleaf), the primary products are anisole and cresols (*ortho-*, *meta-* and *para*-methylphenols), but lesser amounts of xylenols (dimethylphenols) and methylanisoles are also found. Alkylation of both the benzene ring and the oxygen atom is faster when zeolite Y is used, but the selectivity for *ortho*-cresol is greater when ZSM-5 or ZSM-11 is the catalyst.

When phenol is alkylated over ZSM-5 with bulkier reagents, such as propene, propan-1-ol or propan-2-ol, substitution takes place to the exclusion of almost all other reactions; an *ortho/para* ratio of 0.08 is observed (Figure 4.4 overleaf). The formation of bulky disubstituted phenols is also suppressed. It is believed that this is an example of restricted transition-state selectivity.

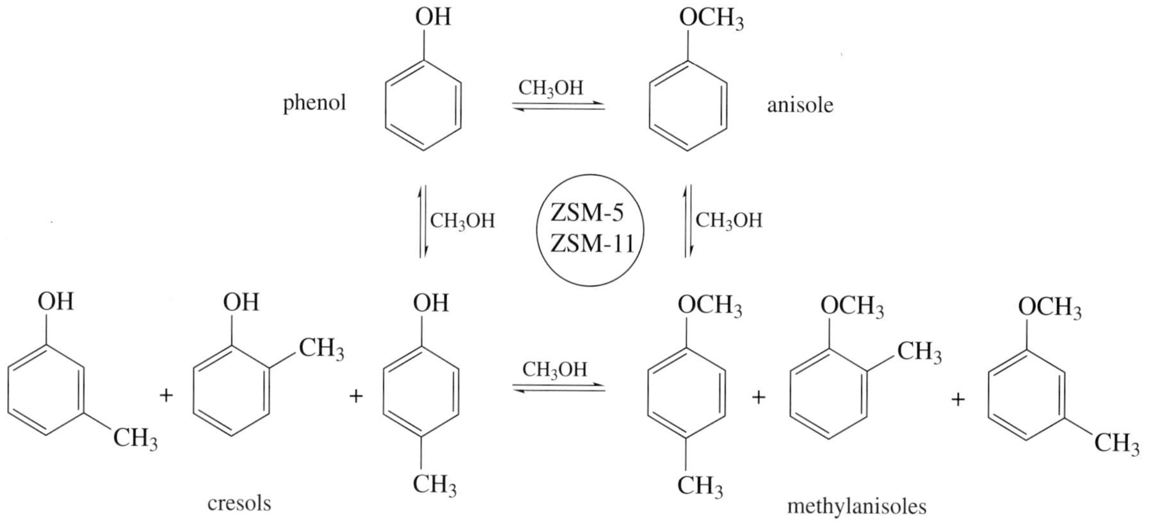

Figure 4.3 Principal reaction pathways for the zeolite-catalysed alkylation of phenol.

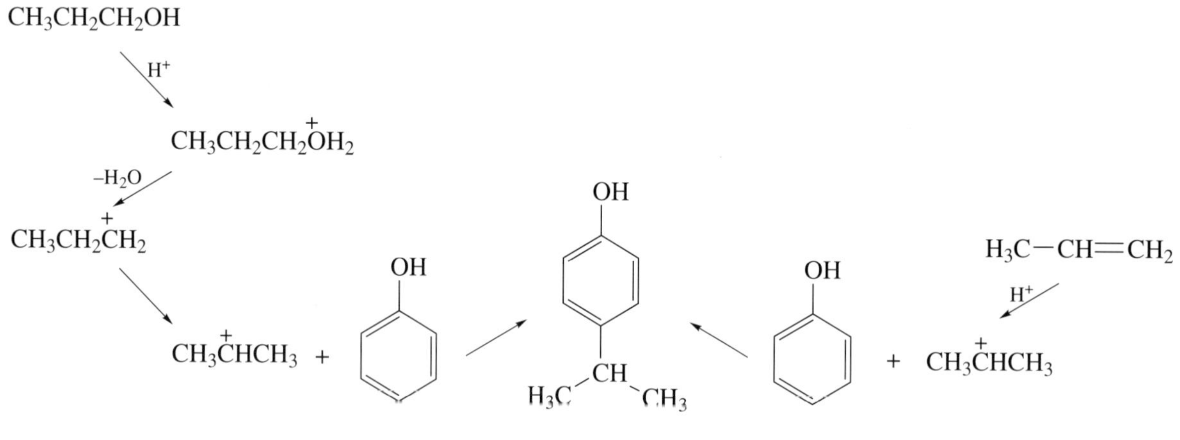

Figure 4.4 The alkylation of phenol with propan-1-ol (or propan-2-ol) and propene.

4.4 Methanol to gasoline

The *methanol to gasoline* (MTG) process uses ZSM-5 to produce good-quality gasoline from methanol. The process can be 'tuned' to yield alkenes for polymer production instead, if these are required. The first plant incorporating MTG technology came on stream in New Zealand in 1986. The overall equation for this process is simple:

$$x CH_3OH = (CH_2)_x + x H_2O$$

But the series of reactions needed to bring about the conversion is much more complex. Indeed, the mechanisms are not yet agreed for all the stages in the process. The first stage is the dehydration of methanol to methoxymethane (dimethyl ether), but there is much controversy over the mechanism of the initial carbon–carbon bond formation. However, once ethene has been formed (presumably via a slow step), well-known fast reactions take over, analogous to those seen when zeolites are used in petroleum refining.

Alkenes will be protonated at acidic sites in the zeolite, and the carbonium ion intermediates will add to the double bonds of neighbouring alkenes, as in the example shown below:

$$2CH_3OH \underset{H_2O}{\overset{-H_2O}{\rightleftharpoons}} CH_3OCH_3 \xrightarrow{-H_2O} C_2 \text{ to } C_5 \text{ alkenes} \longrightarrow \begin{cases} \text{alkanes} \\ \text{cycloalkanes} \\ \text{aromatics} \end{cases}$$

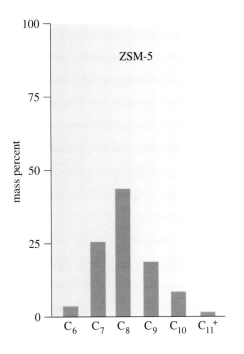

Subsequent reactions will result in further cationic polymerization, cyclization, and aromatization, to form a variety of hydrocarbons. The significant feature is that with ZSM-5, the range of aromatic hydrocarbons is limited to about C_{10}, whereas with mordenite, the range extends up to C_{12} and the distribution is very different (Figure 4.5).

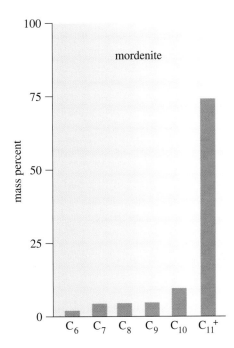

Figure 4.5
The different distributions of aromatic hydrocarbons formed by methanol conversion over ZSM-5 and mordenite, at a pressure of 1 atm (10^5 Pa) and a temperature of 370 °C.

The dimensions of mordenite are sufficient to accommodate molecules as large as hexamethylbenzene (C_{12}), whereas the more restricted pore system of ZSM-5 severely limits the formation of aromatic molecules larger than C_{10}. It is thought that the final stages of alkene condensation take place at the cavities where the pores intersect. Even this space in ZSM-5 is too small to accommodate transition states with more than about 10 carbon atoms. Even if more highly alkylated aromatic hydrocarbons are formed in these spaces, they would find it virtually impossible to escape. So all three shape-selective control processes are operating here – transition-state selectivity at the reaction sites, mass-transport discrimination determining the maximum size of molecules to escape, and molecular traffic control governing the entry of the reagent and the exit of the products.

ZEOLITES AS ENZYME MIMICS

5

The unique catalytic properties of zeolites stem from their efficiency, their specificity, and their stability, under a variety of operating conditions of temperature and pressure, and with a wide range of starting materials. Enzymes are similarly efficient and specific, but their stability under manufacturing conditions is more limited.

In their mode of operation, both zeolites and enzymes begin by bringing reactants and substrates into close proximity via non-chemical binding interactions. In both systems, these interactions have a direct influence on the chemical reaction itself.

The interactions of a reagent with a zeolite are governed by the shape-selective properties and the confinement effects previously mentioned. The cavities and pores of a zeolite can be considered as microreactors, containing stabilized active sites, which can be designed to have particular properties. This is comparable to the active site of an enzyme, where the catalytic activity depends on the three-dimensional arrangement of the functional groups on the side-chains of the amino acids of the protein. Several of these zeolite mimics of enzyme action have been developed, and three are worthy of mention here.

Molybdoenzymes have an active molybdenum atom centre coordinated to sulfur- and oxygen-containing groups in the protein. A variant of zeolite A has been produced with molybdenum atoms coordinated to sulfur in complexes held inside the zeolite cavity.

A haemoglobin analogue has been prepared by using zeolite Y and growing a cobalt complex inside the zeolite cavities. This material is an example of the 'ship in a bottle' synthetic approach, where small fragments are diffused into the zeolite and then encouraged to react to form a complex that is too large to diffuse out of the zeolite. This zeolite has been shown to act as an oxygen binder like haemoglobin.

Another example is the preparation of a mimic of cytochrome P-450, the enzyme responsible for selectively oxidizing organic molecules to hydrophilic compounds, which can either be used in metabolism or excreted. Initially, zeolite A containing palladium and iron (Fe^{2+}) was used, but this very quickly suffered from pore blocking. Then ZSM-5 containing exchanged iron ions was used, using hydrogen peroxide (H_2O_2) as the oxidizing agent. Finally, zeolite Y with cadmium–sulfur clusters was prepared by ion-exchanging cadmium into the zeolite and then treating this with hydrogen sulfide (H_2S) gas.

MESOPOROUS ALUMINOSILICATE STRUCTURES

6

Although zeolite catalysts are valuable for the mediation of reactions taking place in the gaseous or vapour phase, they have a major disadvantage. They have only a limited use in the liquid-phase procedures most frequently used in the manufacture of fine chemicals. Solvent molecules may be adsorbed on to the zeolite surface and so compete with the reagents, in such a way that the ratio of adsorbed reagents is not optimum for reaction. They may also restrict the diffusion of product out of the zeolite, forcing the reaction to slow down, as the chemical equilibrium between the reactants and products shifts back towards the starting materials. The small pores of the zeolites are also susceptible to blockage.

In 1992 scientists from the Mobil Research and Development Corporation developed a new family of **mesoporous** molecular sieves, designated as M41S. Like zeolites, these materials comprise regular arrays of pores, but they have larger pore sizes, in the range 2–10 nm, and they are amorphous rather than crystalline in structure. These mesoporous materials offer much increased diffusion rates in comparison with zeolites, and yet their channels are small enough to provide useful in-pore effects, such as shape selectivity and the local concentration of reagents.

Of particular interest is the material MCM-41, which has a structure consisting of highly ordered, two-dimensional (non-intersecting) hexagonal arrays of very uniform pores (Figure 6.1). MCM-48 is similar, but with a cubic structure. The acronym 'MCM' has never been explained, and the suggestion 'Mobil Crystalline Material' seems unlikely for a near-amorphous solid. MCM-41 is extremely porous, because up to 80% of the 'solid' material can be empty space, and the pore walls have an enormous surface area. A value of twelve hundred square metres *per gram*, an area equivalent to more than a tennis court, would not be unusual!

These mesoporous molecular sieves are prepared using a 'liquid crystal templating' mechanism in which micelles, which are assemblies of cationic alkyl trimethylammonium surfactants $[CH_3(CH_2)_nN^+(CH_3)_3]\,X^-$, act as a template for the formation of the silicaceous material (Figure 6.2). In the silicate-rich aqueous solution, the hydrophobic 'tails' of the surfactant cluster together, leaving the positively charged 'heads' to form the outside of the rod-like 'liquid crystal' micelles. The silicate anions are attracted to, and surround the micelles, aggregating into an open-framework amorphous solid, which precipitates. The solid is filtered off, and heated in air at up to 700 °C (calcination), which removes the surfactant and leaves the

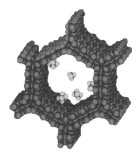

Figure 6.1
Computer representation of methane and ethane molecules inside one of the hexagonal pores of MCM-41. Red = oxygen; blue = silicon; light blue = hydrogen; brown = carbon.

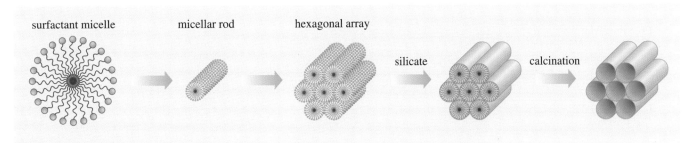

surfactant micelle   micellar rod   hexagonal array   silicate   calcination

Figure 6.2 Preparation of mesoporous MCM-41 by the liquid crystal templating method.

mesoporous silicate. By changing the length of the alkyl chain (the value of n) in the surfactant, the size of the micelle, and hence the diameter of the pores, can be controlled. If the alkyl chains of the surfactant are short, e.g. when $n = 6$, micelles are not formed, and single surfactant molecules are used as a template instead. As a result the medium-pore zeolite ZSM-5 is formed rather than a mesoporous material.

Surprisingly, for a process that seems to be under ionic control, adding appropriate organic molecules to the synthesis medium can bring about an increase in the pore diameter of the resulting material. For example, adding mesitylene (1,3,5-trimethylbenzene) swells the pore diameter by 2 nm, close to the dimensions of the aromatic molecule.

Since their discovery there has been much research and development of ordered mesoporous molecular sieves with high surface areas and adjustable pore diameters. The potential of these materials in catalytic applications is considerable, either as catalysts in their own right or as catalyst supports. The larger pore sizes in M41S and other mesoporous aluminosilicate materials offer opportunities to substitute homogeneous catalytic processes, which require tedious and expensive procedures for isolating and purifying the products, by heterogeneous processes, where the catalyst is simply separated by filtration. Often the catalyst can be reused, either directly, or after a simple reactivation process. Because large volumes of waste and toxic effluent are major concerns for the chemical industry, this prospect has generated considerable interest from researchers working in the areas of 'clean technology' and 'green catalysis' in particular.

The interior of the pores may be modified either by attaching organic molecules to the surfaces or by immobilizing inorganic species there. The large surface areas mean that there are a great number of active sites. The silanol (Si—OH) groups within the silicate structure are able to react with alcohols to form Si—O—C bonds (Figure 6.3a), or with organosilicon compounds to form the stronger Si—O—Si—C bond system. Alternatively, the Si—OH group can be converted into Si—Cl with thionyl chloride, which will then react to form silicon–carbon bonds.

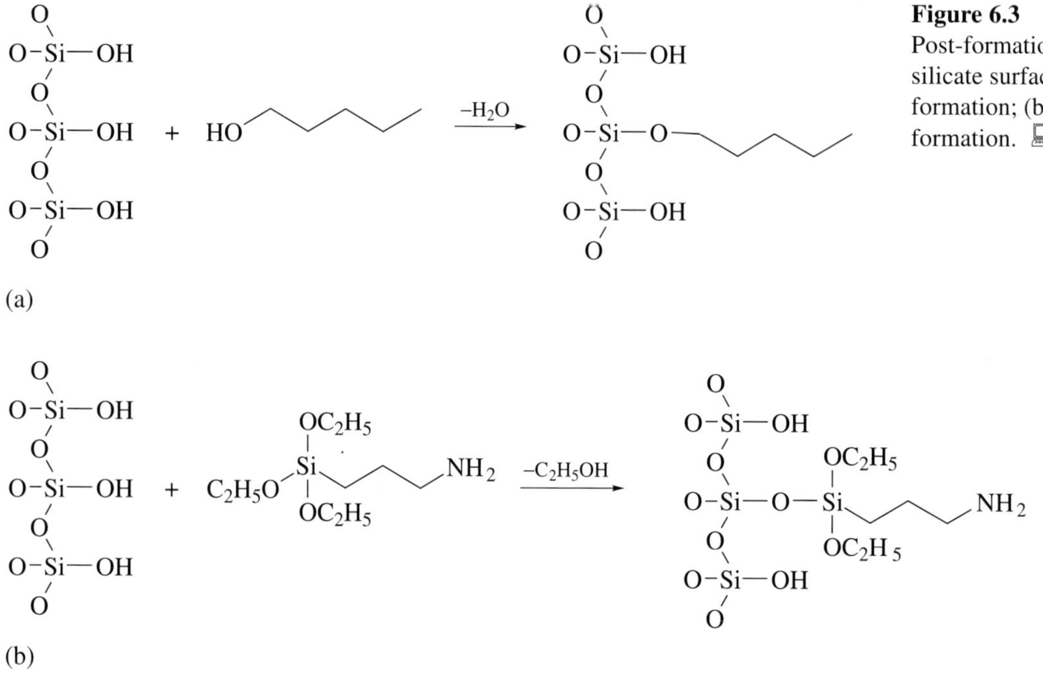

Figure 6.3
Post-formation modification of silicate surfaces: (a) Si—O—C bond formation; (b) Si—O—Si bond formation.

However there is a different way. Just as alumina can be incorporated into the synthesis of zeolites to increase their acidity, so organosilicon compounds, such as organotriethoxysilanes, $RSi(OC_2H_5)_3$, can also be included in the reaction mixture. The organic functional groups introduced this way tend to be more robust than those produced by modification of pre-formed mesoporous structures. If the R group includes a relatively stable functional group, as in aminopropyltriethoxysilane, $H_2NCH_2CH_2CH_2Si(OC_2H_5)_3$ (Figure 6.3b), these functional groups can then be further modified in situ. The unmodified amino group by itself is quite a good *base* catalyst.

Many transition metals and metal complexes have been immobilized within the pores of these mesoporous materials. Partial oxidation of organic materials is a key step in many syntheses, producing a wide range of useful functional groups, including alcohols, carbonyl compounds and epoxides. Traditionally, oxidations are carried out with stoichiometric (or excess) amounts of compounds of toxic transition metals, such as chromium and manganese, in high oxidation states. These reagents are undesirable in both economic and environmental terms. Earlier attempts to use silica-supported oxidants still tended to require stoichiometric amounts, offering only the advantage of easy recovery of the spent oxidant. Increasingly, genuinely catalytic supported oxidants are being developed, by incorporating metals such as titanium and vanadium into silicate structures, and by attaching chromium and manganese to the silicate surface. For example, a titanium analogue of MCM-41, prepared by including titanium tetraethoxide in the reaction mixture, has been used to oxidize alkenes to epoxides. A supported chromium catalyst has also been used to oxidize ethylbenzene to phenylethanone (acetophenone), and a mechanism is suggested in Figure 6.4. Hydrogen peroxide, or even air, can be used as the source of oxygen.

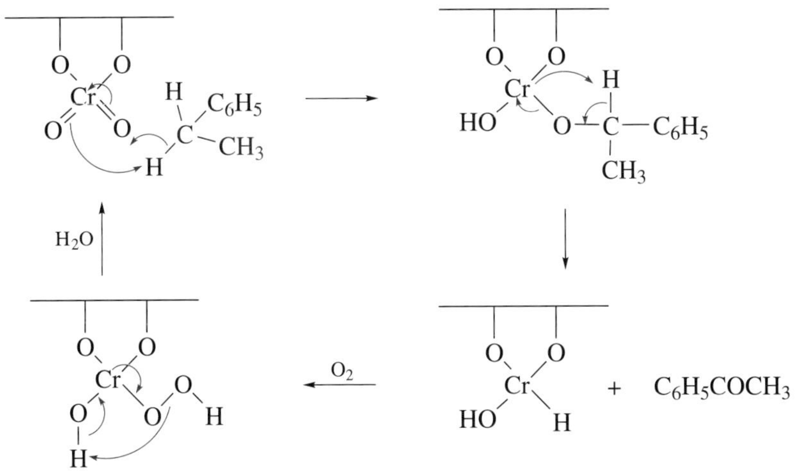

Figure 6.4
Possible mechanism for oxidation catalysed by supported chromium.

Such reactions have also been carried out successfully using aluminium chloride supported on MCM-41, prepared by the reaction of the support with $AlCl_3$. The activity of this Lewis acid is comparable to that of $AlCl_3$ itself, and its selectivity towards monoalkylation is better, because the reaction takes place within the narrow pore structure of the catalyst.

Mesoporous materials have already shown a rich variety of applications. Nickel- and molybdenum-containing MCM-41, for example, has proved to be more effective in the refining of crude oil fractions, and also more efficient at removing sulfur and nitrogen impurities than the materials previously used. They have also been used in polymerization reactions. Poly(aniline), $[-C_6H_4NH-]_n$, a 'molecular wire', or conducting organic material, has been formed inside MCM-41 modified with copper or iron; and a longer chain length is observed when methyl 2-methylpropenoate (methyl methacrylate) is polymerized inside MCM-41, indicating that fewer interactions between growing chains can take place when they are confined in the pores of the material. It is clear that this very active area of research has the potential to yield many more useful discoveries.

CONCLUSION

7

We can see that zeolites have come a long way since they were first introduced as catalysts for petroleum cracking in 1962. The number of catalytic applications where zeolites are now used has risen explosively over the past 15 years. Chemical processes that were established in the first half of the twentieth century may not be considered acceptable by a more environmentally conscious society. Catalytic processes play an increasingly important role in modern industrial chemistry, as the alternatives often produce large amounts of undesirable, and often hazardous waste.

With the advent of environmental awareness in Western Europe, zeolites have come of age. Their high efficiencies and ability to be regenerated repeatedly, the elimination of toxic and hazardous reagents, and the minimization of unwanted by-products, mean that they are an essential ingredient for manufacturers trying to conform to the principles of the Green Chemistry movement. Some zeolite catalysts have been used continuously in processing for over 10 years and have gone through thousands of regeneration cycles.

By using zeolites, the chemical industry has been able to move towards phasing out the use of anhydrous aluminium chloride, one of its most unpleasant reagents, and at the same time to reduce the generation of waste HCl and aluminium salts. The UK no longer manufactures aluminium chloride, and the number of producers in the USA has halved. Aromatic nitration, previously requiring concentrated nitric and sulfuric acids, can now be done over a zeolite using dinitrogen tetroxide (N_2O_4) gas and oxygen from the air, so leaving no acid waste; this use of gases instead of solid or liquid reagents is ideally suited to zeolite catalysis. Another example is the direct formation of amines from ammonia and alkenes, two reagents that would normally not react with each other.

With the advent of mesoporous materials to complement the microporous zeolites, there will be many more developments in the applications of silicate-supported reagents, limited only by the ingenuity and inventiveness of researchers. With increasing pressure from governments for the adoption of more environmentally friendly practices by the chemical industry, the use of framework silicates to provide 'clean' technology can only increase. Zeolites and mesoporous materials offer versatility in application, ease of manufacture, inherent non-toxicity, and convenience of handling, which will ensure their long and varied future, both in industry and in everyday life.

ACKNOWLEDGEMENTS

Grateful acknowledgement is made to the following sources for permission to reproduce material in this book:

Figure 1.1: Amethyst Galleries, Inc.; *Figures 2.4, 3.5*: Nagy, J. B. et.al. (1998) Synthesis, Characterisation and use of Zeolitic Microporous Materials, DecaGen Limited; *Figure 2.7*: International Zeolite Association; *Figures 2.10a, 2.13a, 3.6*: Venuto, P. B. and Landis, P. S. Advances in Catalysis, Vol. 18, Csicsery, S. M. Journal of Catalysis, Vol. 23, Harcourt Brace & Company (Academic); *Figures 2.10b, 2.11b and c, 2.12c, 2.13b*: Meier, W. M. and Olson, D. H., Atlas of Zeolite Structure Types, Butterworth-Heinemann Limited; *Figures 2.10c, 2.12d, 2.13c and d*: Breck, D. W. (1974) Zeolite Molecular Sieves, John Wiley & Sons Limited; *Figures 2.11a, 4.5*: Chang, C. D. (1983) Hydrocarbons from Methanol and Moser, W. R. (ed), (1981) Catalysis or Organic Reactions, Marcel Dekker, Inc.; *Figure 2.12a*: Meier, W. M. (1961) Z. Kristallor, Vol. 115, R. Oldenbourg Verlag, GmbH; *Figure 2.12b*: Reprinted with permission from Journal of the American Chemical Society: Zeolite Chemistry and Catalysis, ACS Monograph Series, Vol. 171, Copyright (1977) American Chemical Society; *Figure 3.1*: Wiesz, P. B. (1980) Pure and Applied Chemistry, Vol. 52. International Union of Pure and Applied Chemistry; *Figure 3.2*: © 1973 IEEE. Reprinted, with permission, from Weisz, P. B. Chemical Technology, Vol. 3, The Institute of Electrical Engineers, Inc.; *Figure 3.3*: Venuto, P. B. (1994) Microporous Materials, Vol. 2. Business Communications Co. Inc.; *Figure 3.4*: British Zeolite Association; *Figure 6.1*: The School of Chemistry, University of Bristol.

Every effort has been made to trace all the copyright owners, but if any has been inadvertently overlooked, the publishers will be pleased to make the necessary arrangements at the first opportunity.

INDEX

Note Principal references are given in bold type; picture and table references are shown in italics.

A

A-factor, 64–5, 69–70
    theoretical, 81–2
accelerated stability studies (drugs), **73–4**
acceleration, 13–15
acetophenone, production of, 253
acidity, and leaving group ability, 157–9, 160
activated complexes, **19**, 20, 83, 84–7, 168
activation control, **84**
activation energy, 64–5, 69, 82
    magnitude of, 71–4
    of reactions in solution, 84
addition reactions, **137**
adsorption, **236**
air bags, 10
alcohols,
    elimination reactions of, 205
    nomenclature of, 139–40
    substitution reactions of, 140–1, 143–4, 159–61
alkenes,
    conversion to epoxides of, 253
    industrial importance of, 190
    synthesis of, 191–2, 194, 195–202, 203–6, 208–11, 212, 248–9
    see also insect pheromones
alkynes, preparation of, 213–14
amides, 195
amines, as nucleophiles, 155–6
amino acids, inversion of enantiomers of, 12
ammonia, as nucleophile, 155–6
anisole; *see* methoxybenzene
anti-**elimination**, **196–7**, 198, 200
antiperiplanar conformation, 196–7, 198, 199
ants, creeping rate of, 75
arctic plants, adaptations of, 75, *76*
aromatic alkylation, 247–8
    para-selective, 246
Arrhenius, Svante, 65, 82

Arrhenius *A*-factor, **64–5**
Arrhenius activation energy, **64–5**
Arrhenius equation, **63–5**
    taking logarithms of, 66
Arrhenius parameters, **64–5**, 81
    determination of, 65–70
Arrhenius plots, **67–70**, *75*
aspartic acid, 12
azomethane, decomposition of, 55

B

Barrer, Richard, 229
bases, **191**
    in elimination reactions, 191–2, 209–10
β-**elimination reactions**, **189–90**, 208–11, 212–14
    mechanisms of, 191–2
        see also E1 mechanism;
            E2 mechanism
Belousov, B. P., 99
benzene, reaction of, with nitric acid, *150*
benzenediazonium chloride, decomposition of, 58
Bernoulli, Daniel, 80
Berthelot, Marcel, 9
bimolecular reactions, 78–9, 209
    elimination, 191
    in gas phase, 80–1
    in solution, 83–4
    substitution, 171
    in zeolite pores, 245
bond heterolysis, **145**
bond homolysis, **145**
bond polarity, 151
bones, dating of fossil, 12
bromide ion, reaction of, with hypochlorite ion, 22–4, *25*, 26, 55–8, 59
bromine,
    reactions of,
        with ethane, 140–1, 145
        with methylcyclohexane, 143–4

bromoalkanes,
    conversion to ethers of, 166
    reactions of,
        with hydroxide ion, 175–6
        with water, 176
bromocyclohexylmethane, synthesis of, 143–4
bromoethane,
    reactions of,
        with ethanol, 172
        with nucleophiles, 163
        with sodium hydroxide, 17, 19
    synthesis of, 140–1
bromomethane, reaction of, with sodium hydroxide, 150–1, *152*, 167, 168–71
2-bromo-2-methylpropane,
    reactions of,
        with ethanol, 172–3
        with sodium hydroxide, 168, 171
1-bromopropane, reaction of, with sodium cyanide, 147
B–Z reaction, 99–100

C

cages (zeolites), **227**
cane sugar; *see* sucrose
carbocations, 17, 20, **171–2**, 173, 203
carbon atoms, classification of, 139–40
cars,
    acceleration of, 13–15
    chemical processes in, 10
catalysts,
    mesoporous materials as, 252–4
    shape-selective, 233, 242–9
    zeolites as, 234–5
cavities (zeolites), **227**
chain reactions, **146**
channels (zeolites), **227**
chemical equations, 16, 17
chemical kinetics, **9**, 11–12
    empirical, 11
    see also rate of change

chlorine,
    reactions of,
        with methane, 145–6, 147
        with propane, 146
chloroalkanes, conversion into iodoalkanes of, 162–3
chloroethane, reaction of, with sodium hydroxide, 136, 137, 147
Christiansen, J. A., 20
cicadas, chirping rate of, 75
classification,
    of carbon atoms, 139–40
    of organic reactions, 136–7
collision theory of reactions, 33–6, **80–1**
completion of reaction, **24**
composite reaction mechanisms, 17–18
composite reactions, 17, 82–3, 91–2
    energy profiles of, 20–21
    see also rate-limiting step
concentration,
    rate of change with time of, 25–9
    reciprocal, 53
    units of, 24
copper, reduction of Cu^{2+}, 91
crocuses, 76
Cronstedt, Baron Axel von, 227, 228
curly arrows, 154
cyclohexylmethanol, reaction of, with hydrogen bromide, 143–4
cytochrome P-450, mimic, 250

D

decarboxylative elimination, 212–13
degree of substitution (and elimination mechanism), **200**
dehalogenation, 212
desiccants, 233
desorption, 236
differential method for experimental rate equations, **45–50**
diffusion control, 84
1,2-dihalides, dehalogenation of, 212, 213–14
dinitrogen pentoxide, decomposition of, 42–5, 51–2, 91–2
drugs, decomposition reactions of, 72–3

E

E alkenes, 196, 198, 205
E1 mechanism, 192, 203–6
E2 mechanism, 191, 192, 194–202, 214
    isomeric alkenes in, 198–202
    scope of, 194–5
    stereochemistry of, 195–7
electrophiles, 152
electrophilic atoms, 152
elementary reactions, 17, 19, 38, 78–88
    collision model of, 33–4
    energy profiles of, 18–19
elimination reactions, 137, **189**
    see also β-elimination reactions
empirical chemical kinetics, 11
enantiomer-selective synthesis, 142
enantiomeric inversion, 12
encounter pairs, 83, 84
encounters, 83, 84
energy barrier to reaction, 19, 20
    see also activation energy
energy profiles, 18–21, 82
enthalpy changes of reaction, 20
enzymes, zeolites mimicking, 250
equilibrium constants (S_N reactions), 161–3, 169
ethane, reaction of, with bromine, 140–1, 145
ethanoic acid, reaction of, with ethanol, 9
ethanol,
    elimination of water from, 137
    reactions of,
        with ethanoic acid, 9
        with halogenoalkanes, 172–3
        with hydrogen bromide, 140–1, 159
ethene, *189*, 190
    reaction of, with hydrogen bromide, 137, 147
    and ripening, 197
    synthesis of, 137
ethoxide ion, reaction of, with iodomethane, 63–4, 67–9
1-ethyl-3-methylbenzene, reaction of, over mordenite, 245
exothermic reactions, 20, 21
experimental rate constant (k_R), **35**, 36
    calculation of, 70
    determination of, 47, 49, 54–5, 58
    effect of temperature on, 71–2

experimental rate equations, 35–6, *37*, 38
    determination of, at fixed temperature, 40–60
        for reactions involving several reactants, 55–60
        for reactions involving single reactant, 42–55
        differential method for, 45–50
        half-life check, 42–5
        integration method for, 50–5
        strategy for, 41–2
    identification of composite reaction from, 90–2

F

faujasite, 239, *240*, 244, 247
femtochemistry, 11, 84–8
first-order processes, 36, 45, 51–2, 79
    E1 reactions, 192
fossils, dating of, 12

G

Gibbs function; *see* standard molar Gibbs function changes
glycotic pathway, 100
gradient of straight line, 13

H

haemoglobin analogue, 250
half-life; *see* reaction half-life
β-halocarboxylic acids, decarboxylation reaction of, 212–13
halogen exchange reactions, 162–3
heartbeat, 100
heterolysis; *see* bond heterolysis
Hofmann, August Wilhelm von, 195
Hofmann elimination reaction, 194
Hofmann's rule, 201
hominids, dating of bones of, 12
homolysis; *see* bond homolysis
Hughes, E. D., 166–7
hydrocarbons,
    cracking of, 234, 242–3
    separation of, on zeolites, 232, 242–4
hydrogen, reaction of, with oxygen, 100
hydrogen bromide,
    reactions of,
        with alcohols, 140–1, 143–4, 159–60
        with ethene, 137, 147

hydrothermal method of zeolite synthesis, **229**

hypochlorite ion,

conversion to chlorate ion of, 94–5, 98

reactions of,

with bromide ion, 22–4, *25*, 26, 55–8, 59

with iodide ion, 91

I

Ingold, C. K., 166–7

initial rate method for experimental rate equations, **55**, 58–60

initial rate of reaction (J_0), **28**–9

insect pheromones, 206

integrated rate equations, **51–5**

integration method for experimental rate equations, **50–5**

intermediate species; *see* reaction intermediates

iodide ion,

reactions of,

with hypochlorite ion, 91

with persulfate ion, 37

iodoalkanes, preparation of, 162–3

iodocyanide, decomposition of, 87

iodomethane,

reactions of,

with ammonia, 155

with ethoxide ion, 63–4, 67–9

ion exchange in zcolites, **233**

ionic reactions, **145**, 147–8

elimination, 191

substitution, 150–64

isolation method for experimental rate equations, **55–8**

K

kelvin (unit), 64

kinetic reaction profiles, **22–4**, *25*, 26–7, 43

kinetic theory of gases, 80

Kinetics Toolkit, 13–14

L

leaving groups, **152–3**

in elimination reactions, 194, 201, 204, 210

in substitution reactions, 156–61, 210

Lindemann, F. A., 20

logarithms, 47

of Arrhenius equation, 66

Lotka, A. J., 99

M

malonic acid, conversion to bromomalonic acid, 99–100

mass-transport discrimination (zeolites), 242–4, 249

Maxwell, James Clerk, 34

Maxwell distribution (molecular speeds), 34–5, 80

MCM-41, 251

analogues of, 253–4

MCM-48, 251

Mellor, J. W., 64

mesoporous materials, **229**, **251–4**, 255

methane, reaction of, with chlorine, 145–6, 147

methanoic acid, reaction of, with sulfuric acid, 99

methanol, conversion to gasoline (MTG process), 245, 248–9

methoxybenzene, alkylation of, 247, *248*

methylcyclohexane, reaction of, with bromine, 143–4

2-methylpropan-2-ol, synthesis of, 141

microporous materials, **227**

molecular beam studies, 81, 85–6

molecular sieving, 229, **232**, 243, 251–4

molecular traffic control, 245, 249

molecularity, **78–9**

molybdoenzymes, mimics of, 250

mordenite, 239, 244, 245, 249

N

nitrogen dioxide, decomposition of, 29, 45–9, 54

nucleophiles, **151**, 152–3, 154–6

and reaction mechanism, 177–8, 209

nucleophilic substitution (S_N) **reactions**, **153–64**

kinetics and mechanism of, 167–73

leaving groups, 156–61, 210

nucleophiles, 151, 152–3, 154–6

nucleophilicity, **163–4**

O

octan-2-ol, synthesis of, 142

order of reaction, 36, 54–5, 79

see also partial orders of reaction

organic reactions, 135–6, 163

classification of, 136–7

oscillating reactions, **99–100**

Ostwald, Wilhelm, 42

overall order of reaction, **36**

oxygen,

reaction of, with hydrogen, 100

see also ozone

oxygen difluoride, decomposition of, 50

ozone, decomposition of, 90–1

P

para-toluenesulfonate esters, 160, 194

partial orders of reaction, **36–8**

fractional, 38

reactions involving several reactants,

by initial rate method, 60

by isolation method, 55–8

reactions involving single reactant,

determination of, 42–55

by differential method, 45–50

by integration method, 50–5

persulfate ion, reaction of, with iodide ion, 37

pH papers, *158*

phenol, alkylation of, 247, *248*

phenylchloromethane, reaction of, with sodium hydroxide, 17–18, 19–20

phenylethyne, *214*

pheromones, 206

photochemistry, 88

physical methods of analysis, **41**

plastics, 190

plausible rate equations, **45–7**, 55–6, 60

poly(aniline), production of, 254

population oscillations, 100

pores, **227**

in mesoporous materials, 252

in zeolites, 227, 232, 235, 237, 238, 239, *240*, 244, 245

potassium *tert*-butoxide, 209–10

pre-equilibrium, **96**

primary carbon atom, 139

products, 136

propane, reaction of, with chlorine, 146

pseudo-order rate constants, **57**, 58

determination of, 60

pseudo-order rate equations, **57**

Q

quartz structure, 230, *231*
quaternary ammonium salts, 194
quenching of reactions, **41**

R

radical reactions, **145–7**
radicals, **145**
rate, 9
rate of change, 9, 12–15
    in chemical kinetics, 22–31
rate of chemical reaction, **29–31**
    effect of temperature on, 63–76
    factors determining, 33–8
    of S_N reactions, 163–4, 175–7
rate equations, 34
    see also experimental rate equations;
        theoretical rate equations
rate-limiting (rate-determining) step,
    94–8
reactants, 136
reaction coordinates (energy profiles), **18**,
    20
reaction half-life, **42–5**
reaction intermediates, **17–18**, 20, 21, 24
    detection of, 90
    femtochemical studies of, 85
    in S_N reactions, 168–70
reaction mechanisms, **16–17**, 90–100,
    139–48
    composite, 17–18
    E1 and E2, 191–2, 194–202, 214
    S_N2 and S_N1, 166–73
reactive sites, 147–8
reagents, **136**
    for elimination/substitution reactions,
        209–10
rhodopsin, 88
Richter, J. B., 16
ripening, 197

S

St Gilles, Péan de, 9
Saytzev's rule, **201**, 204
second-order processes, 36, 46–7, 52–4, 80,
    192
secondary carbon atom, 139
'sentinel effect', 237
shape-selective catalysis, **233**

silanol groups (in zeolites), 233, 235
slope of straight line, 13
S_N1 reactions, 171–3, 175–8
S_N2 reactions, 171, 172–3, 175–8, 195
S_N reactions, **153**
    see also nucleophilic substitution
        reactions; S_N1 reactions; S_N2
        reactions
snowshoe hare, population oscillations of,
    100
sodalite structure, 230, *231*, *232*
sodium azide, 10
sodium cyanide, reaction with
    1-bromopropane, 147
sodium hydroxide,
    reactions of,
        with bromoethane, 17–18
        with bromomethane, 150–1, *152*,
            167, 168–71
        with 2-bromo-2-methylpropane, 168,
            171
        with chloroethane, 137, 147
        with phenylchloromethane, 17–18,
            19–20
'solid acids', 234–5
solutions, reactions in, 83–4
solvent cage, **83**
spectator ions, **151**
standard molar Gibbs function changes, S_N
    reactions, 162
cis-stilbene, isomerization of, 88
stilbite, 228
stoichiometric number, **30**
stoichiometry of reaction, **16**
    1:1, 22
    time-independent, 24
substitution reactions, **137**, 144, 208–11
    ionic, 150–64
    nucleophilic; *see* nucleophilic
        substitution (S_N) reactions
substrates, **136**, 208
sucrose,
    hydrolysis of, 9, *11*
    inversion of, 82
sulfuric acid, reaction of, with methanoic
    acid, 99
***syn*-elimination**, **196–7**

T

temperature,
    effect of,
        on rate of reaction, 63–76
        on theoretical rate constant, 34–5
    and elimination reactions, 210
    reciprocal (inverse), 67
template synthesis, 229, 251–2
terrapin, heartbeat of, 75
tertiary carbon atoms, 139
tetrafluorodiiodoethane, decomposition of,
    87–8
theoretical rate constant (k_{theory}), **34**
theoretical rate equations, **34**, 78–9, 168–
    71
threshold energy (E_0), **80**, 82
time-independent stoichiometry, **24**
transition state, **19**, 20
transition-state selectivity, 245, 247, 249
trialkylbenzenes, production of, 245

U

unimolecular reactions, 78, 171–2, 192,
    209, 211
units,
    A-factor, 65
    concentration, 24
    temperature, 64
    theoretical rate constant, 34
    time, 13

V

van't Hoff, J. H., 45
visual processes, 88

W

water flea (*Alona affinis*), development of,
    74, 75
water softeners, 233
WebLab ViewerLite, 240–1
Wilhelmy, Wilhelm, 9

X

para-xylene, production of, 243–4, 246
xylenes, selective isomerization of, 246–7

Z

Z alkenes, 196, 198

zeolite A (LTA), 230, *232*, 233, 235, 237, 243, 244, 250

zeolite MAP, 229

zeolite X, 230, *232*, 247

zeolite Y, 230, *232*, 247

zeolite ZK4 (LTA), 235

zeolite ZSM-5, 229, 238

    as aromatic alkylation catalyst, 246, 247–8

    as cracking catalyst, 234, 242

    as isomerization catalyst, 243, 247

    as methanol conversion catalyst, 248–9

    molecular traffic control in, 245

    pore sizes in, 244, 245

zeolite ZSM-11, 247, *248*

zeolites, 227–55

    as catalysts, 234–5

        classification of, 235–7

            large-pore, *235*, 239, *240*

            medium-pore, *235*, 238, 242

            small-pore, *235*, 237

    as enzyme mimics, 250

    natural, 227–8

    polarity of, 236–7

    production and uses of, 227

    properties of, 232–3

    shape selectivity of, 242–5

        applications of, 246–9

    structure of, 228, 230–2, 237, 240–1

    synthetic, 228–9

zero-order reactions, 37

Zewail, A. H., 11, 85, 87

Zhabotinsky, A., 99

CD-ROM INFORMATION

Computer specification

The CD-ROMs are designed for use on a PC running Windows 95, 98, ME or 2000. We recommend the following as the minimum hardware specification:

processor	Pentium 400 MHz or compatible
memory (RAM)	32 MB
hard disk free space	100 MB
video resolution	800 × 600 pixels at High Colour (16 bit)
CD-ROM speed	8 × CD-ROM
sound card and speakers	Windows compatible

Computers with higher specification components will provide a smoother presentation of the multimedia materials.

Installing the CD-ROMs

Software, including the *Kinetics Toolkit,* must be installed onto your computer before you can access the applications. Please run INSTALL.EXE from the CD-ROM.

This program may direct you to install other, third party, software applications. You will find the installation programs for these applications in the INSTALL folder on the CD-ROM. To access all the software on the CD-ROM you must install ISIS/Draw and WebLab ViewerLite.

Running the applications on the CD-ROM

You can access the *Chemical kinetics and Mechanism* CD-ROM applications through a CD-ROM Guide (Figure C.1) which is created as part of the installation process. You may open this from the **Start** menu, by selecting **Programs** followed by **The Molecular World**. The CD-ROM Guide has the same title as this book.

The *Kinetics Toolkit* is accessed directly from the **Start | Programs | The Molecular World** menu (Figure C.2).

Problem solving

The contents of this CD-ROM have been through many quality control checks at the Open University, and we do not anticipate that you will encounter difficulties in installing and running the software. However, a website will be maintained at
http://the-molecular-world.open.ac.uk
which records solutions to any faults that are reported to us.

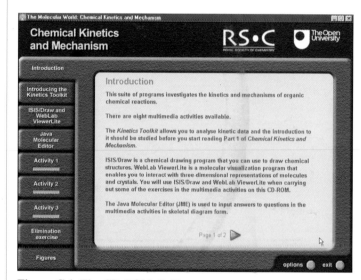

Figure C.1 The CD-ROM Guide.

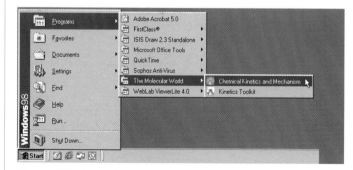

Figure C.2 Accessing the *Data Book* and CD-ROM Guide.